CONSETT IRON
1840 to 1980

Frontispiece. Consett works, 1960

CONSETT IRON

1840 to 1980

A Study in Industrial Location

KENNETH WARREN

CLARENDON PRESS · OXFORD
1990

Oxford University Press, Walton Street, Oxford OX2 6DP
Oxford New York Toronto
Delhi Bombay Calcutta Madras Karachi
Petaling Jaya Singapore Hong Kong Tokyo
Nairobi Dar es Salaam Cape Town
Melbourne Auckland
and associated companies in
Berlin Ibadan

Oxford is a trade mark of Oxford University Press

Published in the United States
by Oxford University Press, New York

© Kenneth Warren, 1990

All rights reserved. No part of this publication may be reproduced, stored in a retrieval system, or transmitted, in any form or by any means, electronic, mechanical, photocopying, recording, or otherwise, without the prior permission of Oxford University Press

This book is sold subject to the condition that it shall not, by way of trade or otherwise, be lent, re-sold, hired out or otherwise circulated without the publisher's prior consent in any form of binding or cover other than that in which it is published and without a similar condition including this condition being imposed on the subsequent purchaser

British Library Cataloguing in Publication Data
Warren, Kenneth, 1931–
Consett Iron, 1840 to 1980: a study in industrial location.
1. North-east England. Iron & steel industries, history
I. Title
338.4'76691'09428
ISBN 0-19-823297-7

Library of Congress Cataloging in Publication Data
Warren, Kenneth.
Consett Iron, 1840–1980: a study in industrial location Kenneth Warren.
Includes bibliographical references.
1. Consett Iron Company—History. 2. Steel industry and trade—Great Britain—History. 3. Iron industry and trade—Great Britain—History. 4. Steel industry and trade—Great Britain—Location—Case studies. 5. Iron industry and trade—Great Britain—Location—Case studies. I. Title.
HD9521.9.C65W37   1990   338.6'042—dc20   89-26508
ISBN 0-19-823297-7

Typeset by Graphicraft Typesetters Ltd, Hong Kong
Printed in Great Britain by
Bookcraft (Bath) Ltd,
Midsomer Norton, Avon

PREFACE

THIS is an account of the complete, 140-year-long career of one of Britain's most distinguished iron and steel companies. Though the focus is on one firm, an attempt has been made throughout to place the development within the context of the economic changes within the industry of the North-East Coast, and to some extent, and particularly over the last sixty years, of those in the nation as a whole.

In writing about Consett Iron I have been fortunate to have had access to the archives of the company up to the mid-1950s. These records are now held by British Steel plc. Records both in the Irthlingborough and Teesside archives have been consulted. At the latter, Mrs E. M. Green has been especially helpful. Events of the more recent decades have also been the subject of conversation with former steelmen both from Consett and from other companies. As these circumstances suggest, the nature of the source material changes over the history of the company. For the earlier years there are company records and an excellent commercial press, but a dearth of statistics, especially of ones which cover the industry as a whole. By contrast, for recent decades there are excellent industry-wide statistics, but the company archives are, as yet, closed to the researcher, and the specialized periodicals, though full of authoritative technical reports, are, for that reason, less useful in terms of commercial intelligence than their Victorian or much weightier early twentieth-century predecessors. The emphasis in the following account inevitably reflects some of these changes.

The book is intended for those within academic fields interested in the study of industrial location, and more broadly in the economic history and economic geography of heavy industry in Britain. Those primarily involved with the processes of regional economic change will, I hope, find it useful as a case study of the role of a dominating firm in such development. Finally, it is my earnest hope that it will interest those in the North-East, and particularly in north-west Durham and in Derwentside, who have a legitimate and fierce pride in a distinguished past and hopes for a future which, in ways still impossible to make out, may yet match its achievements.

K.W.

May 1988

CONTENTS

List of Plates — viii
List of Figures — ix
List of Tables — x
Abbreviations — xiii

Introduction — 1

1. Ironmaking in County Durham and the North-East to 1850 — 5
2. Cleveland Ore and Adjustments at Inland Plants — 11
3. Railway Accommodation for Consett, 1834 to 1862 — 18
4. Consett Crises and Expansion, 1857 to 1870 — 22
5. Early Assessments for Development Planning — 30
6. Re-equipment and the Reorganization of the Supply and Marketing Situation to the mid-1870s — 37
7. The Great Crisis: Collapse in the Finished Iron Trade and Movement into Steel Production — 44
8. The Creation of a Town and a Society—and of the Inertia of Social Overhead Capital — 52
9. Consideration of Relocation in the 1890s — 59
10. Consett, 1890 to 1914 — 67
11. Unravelling the Puzzle of Consett Success — 74
12. The Great War — 83
13. The 1920s: A Critical Decade in Consett Development — 88
14. The Thirties: Depression, and National and Regional Organization — 101
15. Consett Iron and the Redevelopment of Jarrow — 108
16. Consett Works after World War II — 114
17. The New Plate Mill — 125
18. The mid-Sixties: Consett's Location Challenged and Defended — 136
19. Consett within the British Steel Corporation — 145
20. The Progress to Closure, 1980 — 157
21. The Jarrow of the Eighties? — 171
22. One-Industry Communities: Private and Social Costs — 181

References — 186
Index — 190

LIST OF PLATES

Frontispiece. Consett works, 1960	ii
8.1. Consett works and its neighbourhood, 1858	54
13.1. Fell coke works in the early 1950s	91
16.1. Tyne Dock ore terminal, 1957	116
16.2. Consett blast-furnace plant and part of the residential area, 1954	121
16.3. Consett steel plant and part of the town, 1954	123
17.1. Consett works panorama, 1958	126
21.1. The press advertisement of BSC (Industry) after Consett's closure	176

LIST OF FIGURES

1.1.	Consett and neighbouring parts of north-west County Durham, 1827	7
1.2.	Ironworks in the North-East, 1848	8
2.1.	The iron industry of County Durham and Teesside, 1854	13
3.1.	West Durham collieries, ironworks, and railways, 1862	19
4.1.	Distribution of shareholders of the Derwent Iron Company, 1858	24
4.2.	The Consett area in the early 1860s	26–7
8.1.	Consett ironworks and town, 1887	56
9.1.	The Derwent Haugh area, 1895 and 1913	61
16.1.	Consett works, 1954	120
17.1.	Aspects of the operations of the Consett Iron Company in the late 1950s	127
18.1.	Collieries in north-west Durham, 1960	139
18.2.	Consett works in the late 1960s	143
19.1.	Collieries in north-west Durham, 1977	147
20.1.	Consett and total deliveries of billets and billet slabs to various destinations, 1978/1979	164
20.2.	Consett deliveries of slabs, blooms, and billets to other BSC works, 1979	165
20.3.	Steel production of North-East Coast works, 1937, 1945, 1967, and capacity in 1984	169
21.1.	Derwentside industrial estates, 1983	177
22.1.	Consett and neighbouring parts of north-west County Durham, c.1980	185

LIST OF TABLES

2.1. Iron ore and coke costs per ton of pig iron on Teesside and at Consett, c.1860 — 14
2.2. Number of furnaces in Consett, other North-Eastern inland plants, and Teesside ironworks, 1849–1880 — 16
5.1. Pig iron production in Consett, County Durham, and the North Riding — 31
5.2. Coal, coke, and iron production at Consett, 1868 — 31
5.3. Consett capacity for balanced operations, 1869 — 34
5.4. Iron burden at Consett for production of 1,350 tons of pig iron weekly, 1869 — 34
5.5. Ironmaking costs at Consett, 1869 — 34
5.6. Estimated proportions of pig iron sold and used in making finished iron at Consett and in the North-East, 1859, 1866, 1871/2 — 35
6.1. Output of main products by Consett Iron Company, 1865–1871 — 38
6.2. Consumption of Cleveland ore at Consett, 1864–1866 and 1868 — 39
6.3. Deliveries of iron ore to Consett by the North-Eastern Railway, 1868 — 39
6.4. Reductions in freight charges on undamageable iron from Consett, granted by the North-Eastern Railway in 1865 — 40
7.1. Profits in Consett operations, 1870 — 45
7.2. Prices of Bessemer steel and of wrought iron rails and plate, 1873–1881 — 47
7.3. Tonnage of merchant vessels built in and added to the register of the United Kingdom, 1875–1890 — 49
7.4. Consett Iron Company profits, 1870–1887 — 50
7.5. United Kingdom and North-East Coast open hearth production, 1884–1890 — 51
7.6. The condition of North-Eastern ironworks and blast furnaces, 1892 — 51
8.1. Growth of population and housing in the township of Conside cum Knitsley, 1841–1861 — 53
9.1. Estimated savings in weekly shipping costs by using Derwent Haugh, 1896 — 62

List of Tables

9.2. Cost differences for Bessemer pig made at Derwent Haugh for delivery to Consett compared with pig made at Consett	62
9.3. Cost savings in making both iron and steel plate at Derwent Haugh	62
9.4. Estimated land freight costs of assembling materials to make 230,000 tons pig iron, 1896	65
10.1. Gross tonnage of merchant vessels built of steel added to the United Kingdom register, 1880–1913	68
10.2. Consett Iron Company capacity, 1894, 1902/3, and 1911/12	68
10.3. Dividends paid by Consett Iron Company and other North-East Coast iron and steel concerns, 1900–1912	69
10.4. Consett finished steel capacity and the quantity of materials needed to support it, 1905	71
10.5. Coal production by certain outlying collieries, 1895 and 1899	72
11.1. North-East steel shares, 1883	75
11.2. Estimated trends in the structure of productive capacity at Consett Iron, 1894, 1902/3, and 1911/12	77
11.3. Average dividends of leading steel and coal firms, 1898–1910	77
11.4. Shipbuilding in the North-East and other districts, 1889–1920	78
11.5. Production and sales of plate by the members of the North-East Steel Platemakers Association, 1905	79
13.1. Consett Iron Company profits, 1899–1922	89
13.2. Units operated by Consett Iron Company, 1922	92
13.3. Mercantile shipbuilding and ship plate prices in Britain, 1913 and 1919–1929	93
13.4. Shipbuilding and plate production in 1920 and 1921	95
13.5. Dividends on ordinary shares paid by Consett and other North-Eastern iron and steel companies in the 1920s	97
13.6. Profit ratios of North-East Coast steel companies, 1919–1929	97
13.7. Major Consett Iron Company developments in the 1920s	98
14.1. British production of crude and finished steel, 1929–1932	102
14.2. North-Eastern iron, steel, and finished steel production, 1937	107
15.1. Estimated attainable costs for iron and steel made on the North-East Coast and elsewhere, 1935	111
16.1. Crude steel production of Consett Iron and of other North-East Coast firms, 1937, 1943, 1945, and 1950	115
16.2. Plate production by company: North-East, 1943 and 1950; North-East and the rest of the United Kingdom, 1955	122
17.1. Plate capacity, 1955, 1960, and 1965	132

List of Tables

17.2.	Relationship between British shipbuilding and steel production, 1952–1969	133
17.3.	Output of finished products by Consett Iron, 1958 and 1961	133
17.4.	Consett Iron Company, production and financial record, 1956–1966	134
19.1	Number of employees by sector in the district of Derwentside, 1951–1976	146
19.2.	Employment in north-west Durham and Consett, 1956	146
19.3.	1971 forecast of steel production in 1976/7 and 1981/2	151
19.4.	BSC North-East Coast crude steel production in 1980, as envisaged in the 1971 Corporate Development Plan	151
19.5.	Production and consumption of steel plate ($\frac{3}{8}$ in. thick or over), 1970, 1976/7, and 1981/2, as analysed in 1971	151
19.6	The British Steel Industry in 1980/1, as modelled in 1972	152
19.7.	Deliveries of steel plate, 1971–1980	153
19.8.	Consett capacity, 1974 and 1979	154
20.1.	Pig iron and steelmaking at Consett and Teesside works, 1953/4–1981	159
20.2.	Molten iron costs at Consett and other BSC plants, 1978/9	161
20.3.	BSC billet capacity, 1980	162
20.4.	BSC billet and steel capacity and demand, 1979/80 and 1985	163
20.5.	The regional distribution of British demand for steel plate and billets, 1965 and 1980	163
20.6.	The ISTC strategy for Consett semi-finished steel, July 1980	167
21.1.	Employment in County Durham and in Derwentside district, 1951–1979 and 1986	172
21.2.	Population of the Derwentside district, 1951, 1961, 1971, 1976–1983, and 1986	172
21.3.	Population 1961–1986 for districts of County Durham	178
21.4.	Social indicators for districts of County Durham	178
21.5.	Population and unemployment in districts of County Durham, 1985/6	179
21.6.	Unemployment in Derwentside and in the remainder of the Newcastle and Durham travel-to-work areas in January, 1984–1988	180

ABBREVIATIONS

BISF	British Iron and Steel Federation
CDM	Consett Directors Minutes (British Steel NRRC 1027/6/7–8; 701/1/1–5)
BITA	British Iron Trade Association
CFC	Consett Finance Committee (British Steel NRCC 1027/5/2–3)
CG	*Colliery Guardian*
CR	Consett Records (British Steel NRRC 1027): Letter Books and other documents
Econ.	*Economist*
EMRRC	British Steel East Midlands Regional Records Centre
Engin.	*Engineering*
I	*Iron*
ICTR	*Iron and Coal Trades Review*
I and S	*Iron and Steel*
ISI	Iron and Steel Institute
ISTC	Iron and Steel Trades Confederation
JISI	*Journal of the Iron and Steel Institute*
MJ	*Mining Journal*
MSM	*Mining and Smelting Magazine*
NER	North-Eastern Railway
NLR	Newcastle Local Records
NPC	National Ports Council
NRRC	British Steel Northern Regional Records Centre
PCE	*Proceedings of the Institute of Civil Engineers*
PEP	Political and Economic Planning
S and C	*Steel and Coal*
SMT	Securities Management Trust minutes (Bank of England)
ST	*Steel Times*
TES	*Times Engineering Supplement*

Introduction

INDUSTRIAL operations and problems cannot be fully comprehended from balance sheets, although financial statements such as these provide the simplest, least subjective, and therefore, in some ways at least, the most vital indicators of the degree of success which is achieved. Business enterprise must also be set within a variety of other contexts—of competitive conditions, technology, society, and of time and space. In the study of any concern there are general considerations which must be taken into account, but also the insistent pressure of particularity.

In December 1979 the British Steel Corporation gave notice of the ending of steel production at Consett, County Durham. This was merely one instance of a group of widely spread closures, all part of a hurried rationalization programme. BSC had been forced into this by national and international recession and by the replacement of an earlier, politically induced slowness of response to these conditions by a new urgency, which again reflected political pressures, but of a different hue.

Ironmaking began at Consett 140 years earlier. For 130 of those years, the works there had been struggling against the threat of failure in competition with better-located plants. The decisive first step in that long succession of difficult years had been the discovery of the Cleveland Main seam of iron ore in the hills close to Middlesbrough in June 1850. After that, the Teesside iron industry, and later the steel industry which was built upon it, was always an intrinsically lower-cost rival, which could only be kept at bay by an assiduous concentration on superior efficiency in processing, by better marketing, or, and in some respects most important of all, by the pioneering of new processes of manufacture. On the whole the Consett Iron Company had pursued these means with remarkable success, and as a consequence had achieved commercial survival. A slower, but progressive erosion of the position of the Consett works came from fuel economy, marked in ironmaking by reductions in the amount of coke used per ton of iron, and in further processing by the replacement of wrought iron by mild steel between 1875 and 1885. Only at the cost of great efforts was success won in this long competitive race. The achievements of Consett Iron can be seen to have reached a peak in the last two decades of the nineteenth century. This same period was the one in which Britain's steel industry has been accused of falling behind its international rivals, and particularly behind the Americans. Its leaders were said to have been guilty of 'entrepreneurial failure'. If that criticism is justified for the British industry as a whole, then Consett was a marked exception. On the other hand this company may be seen to be an epitome of the opposite conclusion of McCloskey, that 'entrepreneurs in the British iron and steel industry, from whatever perspective they are viewed, performed well' (McCloskey 1973: vii). Less controversially, a by-product of late Victorian success on the part of Consett Iron was the firm establishment of a town whose fortunes were intimately bound up with the continuing activity of the steelworks, and in which the population eventually reached 35,000. In the twentieth century, as social concern became a progressively more important consideration in national locational change, this fact, and the relative isolation of the area, affected decision taking about future developments in the works.

It may be well to say something about the methodological standpoint adopted by the author. Many contemporary social scientists have rejected the belief that academic study can be objective, dispassionate, even-handed in its judgements. It must be acknowledged that each individual is conditioned by family, education, social group, and in turn by his whole value system, but the conclusion that the research worker is thereby invariably committed to either attack or defend the structures of the society which he studies is one which the writer finds unacceptable. Training in an academic discipline involves learning techniques, but additionally, through the gaining of experience, there comes an ability to recognize one's own prejudices and predilections and, as far as possible, to counter them. Commitment to any party or sect should not be the objective either of academic training or of research. The central principle should be that the student is a trenchant critic, guided only by the evidence. Yet even with the best intentions of even-handed treatment one remains only too conscious of personal preferences and the many temptations to bias. It would be too much to hope that one always rejects such temptations.

The respective roles of description, analysis, and theory in an industrial study also deserve comment. Description is often derided—to refer to something as 'mere description' is usually sufficient to condemn it. But description is a necessary preliminary to understanding. Indeed, a description which probes beyond the more obvious features of a situation is itself moving to the understanding of associated circumstances and so towards analysis. Consideration of the debates of boards of directors or the proposals of planners is a prime example of this description which is more than 'mere description'. Statistical analysis may be more impressive, but very often the evidence is not in quantifiable form. Even when it is, the more decisive factors are often human quirks and prejudices—which, as argued above, the researcher should untangle with as much cool objectivity as possible. Theory is a generalization derived from or applied to specific cases. Location theory provides endless fascination for some minds. The writer views it as an invaluable background and as a measuring rod for individual instances of locational practice. Much of what follows concentrates on particularities and details, but as an analysis of locational practice it is also a commentary on the widely known general propositions of conventional, minimum-cost location theory.

Capitalist location theory has its own undoubted logic, but it is a logic which is flouted by numerous examples. Consett was a marked anomaly within the capitalist locational 'game'. The present study is not concerned with the wider social or economic milieu of capitalism, or with the possibilities of a game played according to a different set of economic principles. However, as will be seen, even within a generally free enterprise economy, there were important modifications of the rules of the game as the play proceeded. Some of these involved changes in the relationships between individual players, as with the growth of trade associations and later of industry-wide co-ordination of development. Others, less concrete, were alterations in assumptions as to the purposes of the game.

It is clear that the nature of a capitalist society affects location theory and practice in a variety of ways. On the one hand are assumptions—which are therefore not spelled out at all fully—about the nature of profitability, efficiency, and viability. On the other hand, traditional location theory has not been much concerned with people or with society, except as a source of entrepreneurs, managers, workers, or as the market for the product. At first consideration this restriction of the field may seem justified. One may legitimately deplore the results of locational choice or change as they affect local or regional well-being or the physical and man-made environments, but these are not the main concern of the locational analyst. For him/her the focus is on the maximum efficiency of the production of goods and services. In this context the word 'efficiency' implies the closest possible approach to the point at which the greatest financial return can be obtained for the least input of resources. In short, it covers amoral situations in which private gain by the company is the only interest. Now it is clear that at no time, least of all in the twentieth century, is the private interest of shareholders an adequate object of human striving, but that is the base on which the edifices of locational theory have commonly been built. If one replaces that with the aim of securing greater returns for society as a whole, one may have social justice, but one has certainly lost any pretence to a straightforward theory. Even so it seems to the writer a perfectly legitimate extension of his work to proceed from academic analysis to consideration of public intervention in the locational decisions of private companies, intended to benefit society in specific ways.

Notwithstanding the loss of simplicity and clarity which this approach entails, the career of Consett Iron and the growth of the closely associated town and district provide ample proof of the need for such an interweaving of location theory with social concerns. So far we do not have such a varied base for locational theory in the West, and though we may not wish to adopt models followed elsewhere, the events of the last few decades in western society make clear that this is part of our economic development where hard thought is necessary if we are not to repeat the harrowing experiences of very recent programmes of rushed rationalization of industry. In the final chapters, the writer gives some attention to these issues, but he does not attempt to provide this sort of more comprehensive location theory.

In summary, what follows is an attempt to disentangle some of the elements in the long history of the Consett enterprise, and thus explore the background to the crisis which submerged the works and threatened to destroy the society of the district in the early 1980s. It is a record of a particular, and distinguished, industrial history. At the same time it highlights a number of wider issues, both academic and practical. The history of Consett is in the first instance an outstanding illustration of the fact that conventional location theory is at best a poor guide to locational practice. Locational values are relative and permissive; here, as in other departments of life, difficult circumstances may call forth positive responses, so that even in an inferior location, a plant may have greater success than better-located rivals. Secondly, Consett's history indicates some of the problems which may stem from such a successful reaction to challenging circumstances. A one-industry community is terribly vulnerable, whether in the cut and thrust of a *laissez-faire*, nineteenth-century free market economy, or in an era of rationalizing state industrial monopolies. This leads on to a third theme, a problem for which it must be admitted no solution is offered here. In a period of mature capitalism, when welfare may be recognized as more important than maximum wealth creation, and in which the well-being of the many is normally more highly prized than narrow group interests, should the criteria of the balance sheet and the profit and loss account be the only ones on which the fate of a whole community depends, or should wider considerations outweigh narrowly commercial ones? If the latter, how is success or failure to be assessed and how long will such an enterprise survive in a competitive world? In short, having set out on a business career, can a firm ever escape the commitment to dance to the relentlessly wild rhythm played in the commercial world? It is believed that consideration of Consett's development provides food for thought over fields ranging from theory to policy.

I

Ironmaking in County Durham and the North-East to 1850

THROUGHOUT the first half of the nineteenth century North-East England was a minor factor in the iron industry of Great Britain. In 1847 it made only 5 per cent of the national output of 1,999,000 tons of pig iron. By contrast it was already a major focus of other heavy industries. In the forties it was mining one-third of Britain's coal. It had been, from the beginning, a leader in railway engineering and was one of the main areas for shipbuilding. Taken with other metal processing trades, there was a large and rapidly growing market for iron. The marked lack of response of supply to demand was the outcome of what appeared to be a crippling imbalance in the region's raw material endowment. It had an abundance of excellent coking coal; in the Carboniferous strata west of the coalfield there were inexhaustible and accessible supplies of furnace flux. Unfortunately, as compared with the then leading iron districts, South Wales, the Black Country, and mid-Scotland, iron ore seams in its coal measures were few, poor in quality, and unreliable in occurrence. It is true there were good iron ores in the Carboniferous Limestone—the so-called 'rider' ore, a carbonate of iron of 25 to 40 per cent Fe—but this too was uncertain in occurrence so that no large-scale, sustainable level of production could be ensured. There was also ore in the Scremeston measures of the Limestone Series in Northumberland. Expanding market opportunities eventually brought about attempts to work both these deposits and the scant ore resources of the coalfield. In almost all instances these attempts were plagued by difficulties and proved short-lived.

The smelting of the coalfield ores had been attempted in the eighteenth century, as at Whitehill, near Chester-le-Street, from 1745 to about 1800. By the latter date the two blast furnaces of the Lemington Iron Works, on the Tyne a few miles above Newcastle, were in production. Much of their ore came from pits at Elswick and at Walbottle, but these proved inadequate suppliers, so that the company was forced to haul ore from as far away as Scotland. Such difficulties helped dissuade others from embarking on ironmaking for over a quarter of a century. After that work began on two furnaces at Birtley in 1827. They were active by 1829, and in 1830 made 3,080 tons of pig iron, giving, with Lemington, a total for the North-East of 5,327 tons. The region was then a nonentity in iron. The output of Staffordshire three years before was estimated as 216,000 tons and of South Wales as 272,000 tons. In the thirties Scotland was rising towards national pre-eminence following the application of Neilson's Hot Blast to her Black Band ore and Splint coal. The North-East was still making much less spectacular progress, the old constraint still applying. In 1837 the highly regarded *Penny Cyclopaedia* understandably, but wrongly, informed its readers that 'Clay ironstone is found in Teasdale, but there are no ironworks in this county' (*Penny* 1837). There was greater headway after that; by 1846 there were 35 furnaces in the North-East.[1] There were now blast furnaces in Northumberland—in the interior as at Ridsdale, Hareshaw, and Brinkburn, and on the Tyne at Wylam and at the existing Walker Bar Iron Works. Attention was also given to the development prospects of County Durham.

In 1834 the 30-mile Stanhope and Tyne Railway was opened. The main intention of its promoters was to encourage the expansion of shipments of lime from the quarries around Stanhope through the port of South Shields. However, it also opened up possibilities for other activities in areas of north-west Durham near to its track. Crossing Muggleswick Common from the south-west the railway then ran a mile or two to the east of the river Derwent, crossing the western outcrop of the coal measures. This point of intersection soon appeared an attractive location for a venture into ironmaking (see Fig. 1.1). The translation of a prospect into a development had to await the partly chance arrival of entrepreneurship. The little riverside village of Shotley, known for the settlement there at the end of the seventeenth century of a group of German swordmakers, later also had something of a reputation for the therapeutic quality of its waters. In 1839 William Richardson of Sunderland visited the spa there and became acquainted with John Nicholson, a local mining expert. Together they examined the ironstones of the area. Next, with the co-operation of Robert Wilson of Newcastle, they had bores put down on the plateau above the river near the insignificant settlement then known as Consel or more commonly as Conside. The initial results encouraged further investigations.

The outcome of these mineral evaluations was the coming together in 1840 of four entrepreneurs who agreed to subscribe £10,000 for a new ironmaking enterprise to be named the Derwent Iron Company. Mineral leases were taken up not only at Conside but also to the south-east at the Delves, and near Howne's Gill, the deeply cut channel which continues the line of the middle Derwent through to the headwaters of the upper Browney. Two blast furnaces were erected at first. The business expanded rapidly and by 1845 there were six furnaces with an iron capacity of 400 tons a week. By then 17 puddling furnaces were turning out malleable iron which was rolled down into rails (Louis 1916: 59). By summer 1846 eight blast furnaces were operating and six

---

[1] Porter 1846: 108; *The Times*, 9 Jan. 1864, p. 12.

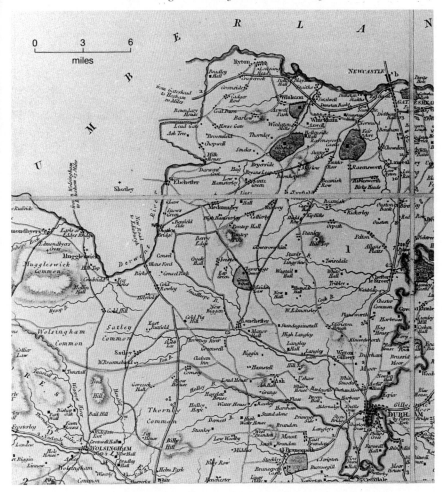

FIG. 1.1. Consett and neighbouring parts of north-west County Durham, 1827

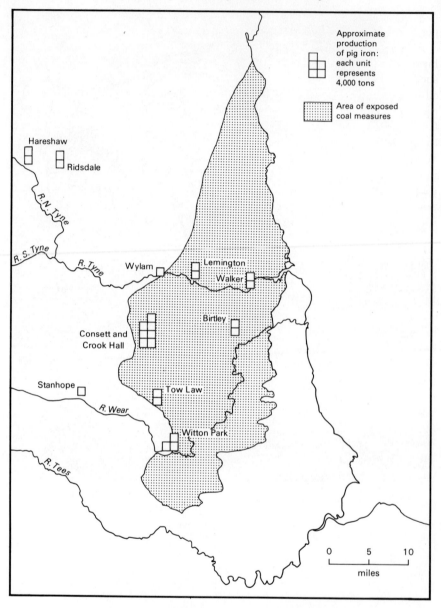

FIG. 1.2. Ironworks in the North-East, 1848

more were about to be lit. In 1848 production of iron was 29,000 tons, or 29 per cent of the total for the region (see Fig. 1.2).

Although at least six Scottish ironworks and eleven in South Wales were bigger, the Derwent Iron Company, judged by growth and size, was already a success.[2] This very achievement exposed it to the problem which had troubled all the promoters or operators of works in the North-East, the unreliability of the ore supply. Early in the history of the plant (exactly when is not known), Cumbrian ore, made accessible by the Newcastle and Carlisle Railway, began to be smelted. Local ore production was also eked out by deliveries of Weardale ore, but good and poor qualities of this were found to be so intermixed that after 7,000 tons of it had been smelted, the practice was discontinued in 1848. More distant regional sources were now explored. Four years before the Derwent Iron works were established the Birtley furnaces had supplemented their failing coal measure ores by obtaining a trial cargo of ironstone from the Esk valley in the North Riding. The new furnaces built at Walker on Tyne between 1842 and 1844 were designed to use Whitby stone mixed with their own mill cinder. In 1849 Derwent Iron turned in the same direction and contracted for deliveries from the Whitby Stone Company and Messrs. Clark of Grosmont.[3]

Notwithstanding its early expansion, difficulties such as these threatened to make Derwent Iron a high-cost producer. The company responded positively by extending their further processing, which gave a higher-value finished product and therefore enabled them to bear somewhat heavier assembly costs than works confined to the sale of pig iron. By 1849 they were already rolling as much as 1,500 tons of bar and flat rolled iron weekly. From this early date can be traced an emphasis on efficiency which was to persist throughout the history of the plant. High furnace productivity was one important factor. Assuming a 300-day operating year, the average daily output of the 24 blast furnaces at work in Durham and Northumberland in 1847 was about 14 tons. By contrast, in 1849 one of the Derwent Iron furnaces produced 218 tons in 6 days, or an average of 36 tons a day, the greatest amount, it was claimed, ever made by one furnace. A correspondent of the time drew the wrong conclusion, suggesting that this 'proves the superior quality of the fuel and ore from the county of Durham'. The importance of the quality of product and excellence of practice was affirmed at a meeting held in the library of the Derwent Iron Company on New Year's Day, 1849. Agents, engineers, foremen, and others met to present their manager, Mr Forster, with 'a testimonial of their esteem'. T. W. Panton, Manager of the subsidiary operation involving puddling furnaces and mills, which Derwent Iron had acquired at Monkwearmouth, Sunderland, stressed that at Consett the ironmaking art was carried to a high pitch—'bar and plate

---

[2] *MJ* 18 Aug. 1849, pp. 397–8.
[3] *MJ* 18 Oct. 1862, p. 720; 8 Sept. 1860, p. 616; 6 Jan. 1849, p. 11; 21 Dec. 1872; 27 Feb. 1875; *CG* 24 Aug. 1879.

iron of every sort and large dimensions, of first rate quality, being produced with an economy of manufacture which, he confessed, surprised him, as it did all who saw it done.'[4] This was, in a sense, merely self-endorsement; but at the same time it was one of the first expressions of a stress on excellence which was to be a recurrent theme throughout the next 130 years.

[4] *MJ* 14 Sept. 1849; 16 June 1849, p. 286; 13 Jan. 1849, p. 14.

2
Cleveland Ore and Adjustments at Inland Plants

WITHIN 10 years of the beginning of the operation at Consett there occured the event which at once rendered it a sub-optimal location for ironmaking. Indeed, from his modern perspective the observer may see its whole subsequent history as in some sense a long-drawn-out rearguard action, although one which was often both glorious and commercially highly successful. On 8 June 1850, the partner in the Witton Park ironworks in south-west Durham, John Vaughan, in company with his mining engineer, John Marley, discovered the outcrop of the Cleveland Main Seam on the slopes of Eston Hill, overlooking the Tees estuary. This solved the general problem of north-eastern iron ore supply, but at the same time threatened the extinction of those furnaces in the interior of the region, which were often already struggling. The North-East Coast had suddenly been transformed from being poorly endowed for ironmaking to being superbly provided for it, but locationally the regional situation had been turned inside out.

Works such as Tow Law and Consett, established on or near to local supplies of both fuel and ore, were now, as it was quickly recognized, disadvantaged as compared with the new furnace plants that were soon going up, one after another, along the Teesside flats.[1] Cleveland ore as quarried or mined ran 30 per cent iron or little more, and even when calcined (that is, when its moisture was driven off), its Fe content was only about 40 per cent, so that from 2.5 to over 3 tons of ore were needed for every ton of pig iron. The employment of this low-grade ore resulted in an increase in the locational attractions of the south-eastern parts of the district. However, the best coking coals were in the south-west and the north-west of Durham, and furnace limestone came from quarries lying beyond the coalfield in the Carboniferous Limestone of Teesdale and Weardale. The markets for both pig iron and finished iron products lay back on the estuaries, in other parts of Britain, or overseas. Most North-East Coast works specialized in iron rails. Consett too was involved in this trade, but it was also an important producer of plate iron—of which indeed it was, until into the sixties, the only large-scale source within the region. From the early 1850s large tonnages of Consett rolled

[1] *CG* 16 Jan. 1858, p. 40.

iron were being delivered to Clydeside shipbuilders. A few years later the Admiralty yard at Chatham was using Consett plate and angle iron.[2]

The locational situation was more or less in a state of balance as far as the respective attractions of Teesside and the coalfield were concerned in the early 1850s; together, fuel economy and the growing importance of Cleveland ore could only swing it more and more in favour of the coastal districts. This was recognized in the evidence which the proprietor of the Weardale Iron Company, Charles Attwood, gave in 1861 to the House of Commons Committee on Durham Railway Bills. He described the situation at Tow Law, 8 miles south of Consett, and similarly placed in respect of materials and markets. Each year they made about 40,000 tons of iron. For this it was necessary to assemble 70,000 tons of coal and 60,000 tons of coke—though he put the coal equivalence of the fuel used for iron smelting alone at about 90,000 tons. Each ton of iron required half a ton of limestone. If all the ore smelted at Tow Law had come from Cleveland they would have required some 120,000 tons of it. In Attwood's case there was an eastward movement of operations marked by the building of works at Tudhoe. Even earlier Bolckow and Vaughan had supplemented the ironmaking capacity of Witton Park with blast furnaces at Eston and Middlesbrough (see Fig. 2.1).

Though they recognized the advantages it possessed in having good coal, cheaply worked, contemporaries also saw that Consett's position had been made relatively more difficult. One report noted that the newly opened Stockton Malleable Iron Company was managed by men formerly with Consett, 'which *en passant*, I may say are not likely to be improved by the opening of ironworks on the Tees'.[3] In the same railway bill hearings to which Attwood explained the position of his company, a Mr Johnston observed, 'At Conside, they are under the *fatal* necessity of using large quantities of Cleveland ironstone, because they have been *foolishly* established in a situation where there was a very small amount of ironstone, which was speedily exhausted.'[4] Derwent Iron seemed to give substance to these critical opinions of its worsened circumstances by acting very quickly to acquire control of Cleveland ore supplies. Within less than seven months of the discovery of the Main Seam, they had an agreement with the Pease company for 10 years supply of from 8,000 to 10,000 tons of Cleveland ore a month at 8*s*. 3*d*. a ton.[5] A little later they leased ore lands at Upleatham from Lord Zetland. By July 1852 a railway was being built from the Middlesbrough to Redcar line to open up these deposits. The ore property was a good one, the Main Seam being undivided, though only 13 feet thick as compared with 15 feet 2 inches at Eston (Bird 1881: 120–1). The outcome of these investments and develop-

[2] *MJ* 7 Dec. 1850, p. 585, 4 June 1853, p. 340.
[3] *CG* 8 June 1861, p. 358; *MJ* 15 Jan. 1859, p. 49, 30 Aug. 1862, p. 595.
[4] *CG* 8 June 1861, p. 358 (author's italics).
[5] David Dale papers (NRRC 1027/3/31–2).

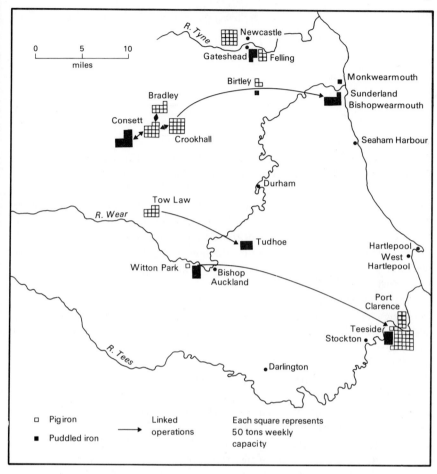

FIG. 2.1. The iron industry of County Durham and Teesside, 1854

ments was to kill off the remnants of coal measure ore production in the Consett area. At one time they had been using local clayband iron ore at 5s. a ton, at 'a ruinous sacrifice' as it was put, when Bolckow's Eston works were paying only 2s. 6d. for ore at the furnace. In the mid-50s coal measure ore was 10s. a ton delivered at Consett furnaces, when Cleveland ore could be mined, carried via Bishop Auckland, and delivered to Consett for 7s. a ton. In 1856 the situation in favour of the use of Cleveland ore was improved still further as a result of a new agreement on ore carriage with the Stockton and Darlington Railway. By 1857 Derwent Iron had closed all its ore workings in the Shotley Bridge area, though occasionally in the future some local ore was mixed with the Cleveland stone to produce irons of special qualities.

Employment of Cleveland ore, though commercially attractive, was not

TABLE 2.1. *Iron ore and coke costs per ton of pig iron on Teesside and at Consett c.1860*

	Cost of Cleveland ore (3 tons)		Cost of Durham coke (2 tons)	
	s.	d.	s.	d.
Eston	10	6	20	0
Consett	21	9	16	0

Source: MSM 1862/3.

without its problems. In the first place its delivery to Consett was not, as yet, straightforward. The journey to the plateau on which the Consett furnaces stood had to be completed by rope incline. Secondly, there is clear evidence that the Upleatham mining venture was not well conducted. The mines were opened on the 'wrong side' of the hill, which meant that there was a break of bulk at Redcar. In 1857 control of these mines passed from Derwent Iron to Pease and Partners. At that time production was only 58,000 tons. The new owners developed the property with much greater vigour, so that within eight years output reached 718,000 tons.

Meanwhile doubts continued about Consett's long term survival prospects in the new circumstances of regional iron production.[6] In the early sixties the situation was surveyed in admirable detail (though with some inaccuracies in the apparent precision with which costs were quoted) by two eminent French metallurgists, M. Grunner and M. Lan. They visited the main British metal districts and their impressions were reported very fully in the *Annales des mines*, and summarized in the British trade journals. Their conclusion was that the raw material supply situation in the North-East was not particularly unfavourable to Derwent Iron. Coke at the ovens was 8s. a ton, or 1s. more for the better grades. It could be carried some 30 miles to Middlesbrough for 2s. a ton. At 2 tons of coke per ton of pig iron, Consett had costs which on the fuel transport account were of the order of 4s. a ton lower than Teesside. Cleveland ore they priced at 3s. 3d. a ton loaded on wagons. (A low figure compared with the price at which Bolckows were then selling the ore which they did not require themselves.) Consett was paying 3s. 9d. a ton for the carriage of ore from there. They were at the time again using a small amount of local ore at 11s. a ton, and some 66 per cent Fe Cumberland ore at 20s. a ton, including 8s. for carriage (see Table 2.1). In spite of its apparent greater transport charges, Consett works was, they found, able to keep overall costs almost competitive with those on Teesside, presumably by achieving lower process costs. But, adding freight charges for carriage to the ports, some 4s. 6d. a ton, Grunner and Lan naturally concluded that Consett could not compete successfully in the export trade in pig iron. They stressed what had already become obvious in

---

[6] *MJ* 18 Oct. 1862, p. 720; Fordyce 1860: 149; *MJ* 15 Jan. 1859, p. 49, 12 Mar. 1859, p. 179; *CG* 10 Nov. 1876, p. 747; *MJ* 18 Oct. 1862, p. 720.

Derwent Iron's history—the need to develop the finishing lines as a way out of these difficulties of location. Out of an 1859 make of 80,000 to 90,000 tons of iron—only about two-thirds of capacity—25,000 tons went into rails, 13,000 tons into plate, and 4,000 tons into bars.[7] Over the next few years finished-iron production was half or more the tonnage of pig iron.

Already, in addition to its emphasis on further processing, on quality, and on efficiency, Consett had other advantages. Its size gave it substantial bargaining power with those who sold it raw materials and services. In 1855 with 14 blast furnaces it was still the biggest ironworks in the region. By contrast, at the same time Bolckow and Vaughan's 13 furnaces were divided between the three separate locations of Eston, Middlesbrough, and Witton Park. Moreover, Consett was a major source of business in a district whose trade it dominated. Already Derwent Iron paid about £150,000 a year for rail carriage. It used the negotiating power thus acquired to obtain competitive freight rates, and by that means to open better sources of supply. The North-Eastern system had charged it 3s. 9d. to carry iron to the docks at Jarrow, but by 1859/60 Derwent Iron had obtained a reduction to 3s. 2d. (Fordyce 1860: 151). At the same time an agreement was drawn up for Consett to take 160,000 tons annually for 17 years of the magnetite ore from Rosedale, in the North York Moors, at 6s. 10d. a ton. In contrast with Cleveland stone, this ore ran at 49 per cent Fe, so that only 2 tons were needed per ton of iron, and price at the mine and transport cost together would substantially cut ironmaking costs on the ore account. Cumbrian ore remained of far greater significance than was the case generally in the North-East. In 1862 the North-Eastern delivered 38,445 tons of Whitehaven ore to Derwent Iron, but only 16,647 tons to all the other works in the region. In fact, despite its early involvement in the Cleveland orefield, by the mid-1860s Consett was only to a small extent dependent on it. In 1864 and 1865 it used 111,000 tons of Cleveland ore. At 30 per cent, this is equivalent to only 33,000 tons of pig iron. In the same two years its iron production totalled 179,000 tons. In 1868 83.5 per cent of the ore delivered to Consett by the NER came from the North-West. Given that this ore was twice as rich as the normal run of Cleveland stone, operations there at that time may be regarded as being essentially dependent on west-coast haematites.

Elsewhere in the region the works in the interior had less success. Ridsdale and Hareshaw, both two years older than Consett, failed in 1857. At the former almost all the capital of nearly £170,000 was lost. Hareshaw's costs of production and delivery were said to be about £4 a ton when the selling price was £3. By the time the inevitable collapse came, the company owed the Union Bank some £152,000.[8] In other inland plants which survived, the utilization rates were both then and afterwards in general noticeably poorer than those on Teesside and at Consett (see Table 2.2).

[7] *MSM* May/June 1863, pp. 324–32.   [8] Bell papers.

TABLE 2.2. *Number of furnaces in Consett, other North-Eastern inland plants, and Teesside ironworks, 1849–1880*

	1849	1856		1862		1869		1871		1880	
	Inblast	Existing	Inblast	Existing	Inblast	Existing	Inblast	Existing	Inblast	Existing	Inblast
Teesside[a]	none	29	27	40	33	79	62	79	77	110	91
Consett	14	14	14	18	4[b]	17	5	6	6	7	6
Other interior locations[c]	20	28	22	29	14	43[d]	19[d]	33	28	35	8

[a] At or below Stockton and within easy reach of tidewater.
[b] Consett was at this time in financial crisis and reconstruction.
[c] Works judged not accessible to water carriage of iron ore.
[d] It is assumed that all the Bolckow Vaughan Witton Park furnaces were active.

In the same year as Ridsdale and Hareshaw collapsed, the Derwent Iron Company also ran into a financial crisis. In this case however the causes were largely external, and a combination of vigorous management and the recognition of the company's importance to the district economy ensured that it remained in production. Indeed, in a number of important respects Consett was to gain lasting advantages from the problems which then for a short time threatened to overwhelm it.

3
Railway Accommodation for Consett, 1834 to 1862

NOT surprisingly, Derwent Iron was not only vigilant in the quest for favourable freight rates, but also active in promoting the physical provision of railway services in its section of north-west Durham. Over a period of less than 30 years, the high and isolated plateau on which the works were built was provided with important rail links in five directions: south-westwards to the Carboniferous Limestone of Weardale; to the markets and the export points of the lower Tyne; via the Derwent valley to the industrial districts of upper and middle Tyneside; down the Lanchester valley to make connection with the east coast main line; and south-east across the western section of the Durham coalfield to Bishop Auckland, to Teesside, and so through to the iron ores of the Cleveland Hills (see Fig. 3.1).

The Stanhope and Tyne Railway, opened six years before construction of the ironworks began, provided the latter with its first means of bulk transport. In 1843 a short link line was built to the rival Brandling Junction Railway, so giving Derwent Iron direct access to Gateshead. By this time the company was also looking southwards. It gained a first interest in this direction through the business problems of others. The Stanhope and Tyne Railway Company was in difficulties by 1841 and was then incorporated in the new Durham Junction Railway. In the following year the upper parts of the old Stanhope and Tyne system—essentially the section above Consett—along with the Stanhope quarries were sold to Derwent Iron. These developments brought operations in the Consett area into relations with the Stockton and Darlington Railway Company which was expanding from the south of County Durham.

The board of the Stockton and Darlington had, from the beginning, entertained the idea of extending their system from the coalfield into the limestone districts to the west. By 1843 the need to meet competition caused them to build a line north to Crook. Derwent Iron now projected a railway from the newly acquired western section of the Stanhope and Tyne to meet the Stockton and Darlington at Crook. However, in 1845 they leased both their existing tracks and the Stanhope limestone quarries to the Stockton and Darlington, committing themselves at the same time to provide them with substantial business. In the same peak year of national railway construction mania, the Stockton and Darlington opened their new line from Crook to the

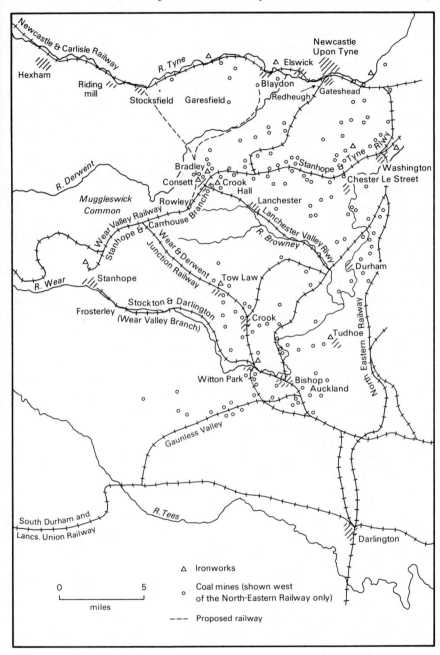

FIG. 3.1. West Durham collieries, ironworks, and railways, 1862. Only main railways are shown

old Stanhope and Tyne and so through to Consett in the form of the Wear and Derwent Junction Railway. Six years later, when Derwent Iron opened iron mines in Cleveland, the existence of this railway connection enabled the Stockton and Darlington to carry the ore 54 miles direct to Consett, though for some years not without the inconvenience and extra costs of rope incline movement for part of the journey. (In 1853, for instance, up to 1,000 tons of ore a day were roped up the Howden incline on the Crook section of the route.)

Utilization of Cleveland iron ore at Consett stimulated two other railway projects. As shown above, in the early days it was found desirable to mix some haematite ore with the Cleveland stone, not only at Consett, but also, though to a smaller extent, on Teesside. The Newcastle and Carlisle Railway delivered Cumbrian ore from the west coast to Redheugh on the Tyne, but in 1856 the Stockton and Darlington moved to gain a share of this trade by forming the Stockton and Darlington and Newcastle and Carlisle Union railway to construct a line from Rowley on the Stanhope and Tyne to meet the Newcastle and Carlisle at Stocksfield. Work was begun early in 1857 with the expectation that by means of the new line not only would west coast ore come directly to Consett, but it would go on via the Wear and Derwent Junction to Tow Law and to Middlesbrough. However, this project was cut in the bud by two blows when the only part which had been built was the short stretch between Consett and Crook Hall.

The first blow was that in autumn 1856 plans were produced for a more direct line to the west coast from Hagger Leases in the Gaunless valley west of Bishop Auckland to Kirkby Stephen and Tebay. This railway was sanctioned in 1857 and opened in 1861. The second blow came in November 1857, when the suspension of payments by the Northumberland and Durham District Bank brought with it the failure of one of its largest customers, the Derwent Iron Company. The Rowley to Stocksfield link was never completed. A few years after this two other important lines were built. The expanding new railway consolidation, the North-Eastern Railway, wishing to gain a share in the now major traffic in Cleveland ore, in competition with the Stockton and Darlington, obtained approval in 1857 for a track up the Browney valley from their main line and on to Consett. This Lanchester Valley Railway was opened in 1862. In 1861 there were plans for another link from the Newcastle and Carlisle to Consett: this became the Derwent Valley Railway from Blaydon, opened for mineral traffic in 1867.

In spite of the number of initially independent, competing rail routes to Consett, eventually all of them were controlled by the North-Eastern. In February 1859 this company approved terms for union with the Newcastle and Carlisle Railway, and after some difficulties, the union was completed in 1862. The Stockton and Darlington was merged into the North-Eastern in the following year. Consett was henceforth to be served by only one railway system: the size of its operations meant that, even so, assiduous attention to its needs could never rationally be ignored by North-Eastern managers and directors.

The care with which, from an early stage, Derwent Iron pursued its search for favourable treatment by the railway companies, in order to counteract any tendency for its operations to become marginal, may be appreciated by considering the pressures it placed on the Stockton and Darlington at the time when the competition of Teesside ironworks was first becoming serious. The board minutes of the railway company are revealing. On 10 January 1853 a special meeting of the Stockton and Darlington board was called to consider Derwent Iron proposals for the modification of the 'Ironstone Leading Contract'. They proposed that they should increase the amount of ore which they guaranteed to take annually to 350,000 tons in return for rate reductions by the railway company. On this occasion the iron company was to be disappointed. The Stockton and Darlington minutes recorded that their board 'has fully deliberated upon said proposition and after weighing all the circumstances, both as to cost of additional stock at highly increased prices and the general requirements of such a trade, it is resolved that the Secretary be instructed to inform the parties that the Company cannot agree to make an abatement from their present low rate of charge'. The struggle continued under different forms, Derwent Iron often winning concessions. By July in the same year, when the supply of limestone from Stanhope quarries to the furnaces at Crook Hall and Consett was falling behind requirements, the Railway's inspector was authorized to arrange that the traffic be increased to over 100 wagons a day. In September the Engineer of the Stockton and Darlington was reported to have produced plans for a permanent structure across the deep valley of Hownes Gill, so easing a difficult stage of the delivery. A few months later, he was directed to examine ways of improving the links between Hownes Gill and the Wear Valley line extension so as to avoid the difficult stretch known as Myer's Incline.[1]

The material improvements to the railway system and the never-failing vigilance of the management at Consett in their search for good and, if possible, cheap railway service helped to reduce the liabilities associated with their plant's inland location. At the same time emphasis continued on the importance of efficient operations at the furnaces in order to minimize the contribution which process costs made to overall costs of production. In the late fifties and early sixties these preoccupations were interrupted by threats to the continuation of the works which came from a completely different quarter.

---

[1] Stockton and Darlington Railway, board minutes 10 Jan. 1853, 22 July 1853, 9 Sept. 1853, 10 Mar. 1854.

4
Consett Crises and Expansion, 1857 to 1870

TEN years after its establishment Consett faced the challenge which stemmed from the discovery of the main section of the Cleveland ore. Seven years after that, it passed through another quite different sort of trauma. In November 1857 the Northumberland and Durham District Bank failed. The Derwent Iron Company owed it almost one million pounds, and therefore was also faced with collapse. There was some dispute as to whether or not the ironworks should be saved, and this debate raised some of the wider issues which were to be considered again and again throughout the long history of ironmaking at Consett.

Early in the crisis it seemed that the Northumberland and Durham District Bank was not prepared to wait for the sums due from Derwent Iron. A discussion of policy resulted among the Quakers who were so prominent in the affairs of that concern. One of them, William Backhouse, outlined the pressures the bank was placing upon them: 'we are to assign all we have to them and they are to be at liberty to take any proceedings against the company.' In his efforts to save the iron company, Backhouse asked Johnathan Pease to meet him in Darlington or Newcastle, 'or would any of you like to come here first and look at our capabilities, costs of producing pigs etc, etc'. Pease replied with what was effectively a recognition of the wider, multiplier effects of such a great enterprise as Consett had already become. 'After stating my continued sympathy, I wish to tell thee that meditating upon your position I am quite of the mind that it is most clearly to the interest of your proprietary, the Bank and the Poor as well as My Brother and the two Railway Companies that you should if possible continue your works with as little change or interruption as possible.' Meanwhile, in a printed statement circulated to shareholders in the Northumberland and Durham Bank, Jonathan Richardson, another of those with important interests, took a different view. He argued that when Derwent Iron fell into debt, they could and should have closed the works. But this he acknowledged would have damaged the bank, and he also suggested, not very convincingly, that it had been felt that the discovery of the Cleveland stone, 'which has produced such important results elsewhere, would have enabled the Iron company to surmount their difficulties'. Backhouse was 'stung to the very quick' by the attitude which Richardson had taken, and by the fact that possession had been taken of the works. For a time it seemed that pressure from Scottish and other banks would bring about a liquidation. Under this new, crisis period of control, one 'valued servant' was quickly dismissed, and of those who gained influence and had rival interests, it was said that they 'will not hesitate to sacrifice Consett'. Others too foresaw the dire consequences of a failure. As a leading trade journal put it in January 1858, 'The stoppage of the Derwent Ironworks would have been a calamity not only to the immediate district, but to the whole district.' In fact at this time in its history the operations of Derwent Iron, in coal-mining, coking, ironmaking and iron finishing, were already said to be effectively the sole support for a population of 30,000.

Gradually a way was found to a solution that preserved the works. In February Pease wrote to the liquidators at the Northumberland and Durham Bank giving his opinions of 'the general prospects of the exceedingly important business committed to your charge, having especial reference to the maintenance of the concerns of the Derwent Iron Company in complete working order'. By the middle of the same month Charles Bragg reported to Pease his own assessment of the situation: 'I have gone carefully thro' the whole of the manufacturing costs and statistics, and I am convinced that with 5% being only charged upon the overdrawn capital and an average run of commercial prosperity Consett has the elements, with close and prudent business care, to produce profits ample to enable Jonathan Richardson to fulfill his guarantee.' At the end of the month, Richardson, apparently still sceptical, wrote on behalf of the lawyers representing the Scottish banks and the liquidators to the young David Dale: 'I think they must know the chances of carrying on the works before they can be expected to make or authorize any sacrifices in order to accomplish that.' More cheerfully he added, 'I am very hopeful now that your persevering efforts to make the best of a bad job will not be in vain.'[1]

An apparent solution to the problem was found when 54 shareholders of the Northumberland and Durham District Bank, incorporated as the Derwent and Consett Iron Company Ltd., acquired all the properties of the Derwent Iron Company for £930,000. More than half of the shareholders in the new concern lived in the North-East (see Fig. 4.1). The Priestman family became one of the largest holders of stock, and Jonathan Priestman became joint manager with David Dale, who had been appointed to the group conducting the operations of the ironworks to represent the interests of the Cleveland ore line, the Middlesbrough and Guisborough Railway.[2] The railways also responded financially to the difficulties, as well they might to a concern which now assembled each year some 0.5 million tons of ore and limestone and 0.6

---

[1] The correspondence concerning the crisis, and covering the period Nov. 1857 to Feb. 1858, is in the Derwent papers, D/HO/F119, Durham County Record Office. See also *CG* 30 Jan. 1858, 14 Sept. 1861.
[2] *NLR* 21 Dec. 1888, Obituary of J. Priestman.

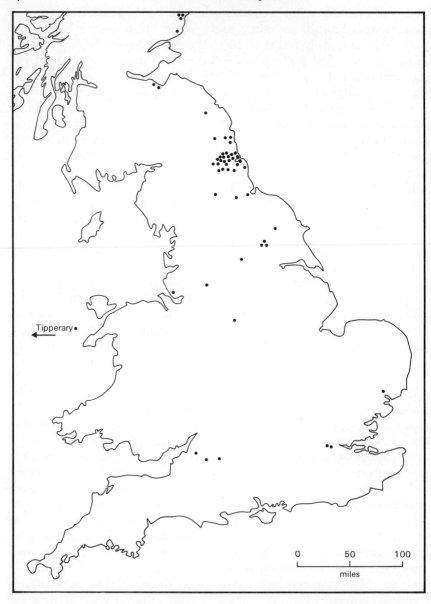

FIG. 4.1. Distribution of shareholders of the Derwent Iron Company, 1858. One shareholder location (Samieston House) is unidentified and could not be placed

million tons of coal, and which dispatched 0.15 million tons of finished products. One comment in particular struck home: 'the stoppage of the works would inflict such an evil on the district ... that such a thing is not to be thought of for a moment.' Fortunately the railways did think of it, and acted accordingly in their own interest. The North-Eastern Railway provided credits of £200,000.[3]

Although, despite depression of trade, the profits of the new concern over the next six years averaged 12 per cent, the reorganization of 1858 did not prove a success. The newly formed company was unable to complete the purchase. A second reconstruction produced a company which began to trade in April 1864 as the Consett Iron Company Ltd. Consett Iron acquired extensive coal and coke interests, and, at the three neighbouring works of Consett, Crookhall, and Bradley, 18 blast furnaces, of which only six were then at work. Consett works had 99 puddling furnaces and there were 31 more at Bishopwearmouth. The annual capacity was reported as 150,000 tons of pig iron and 50,000 tons of finished iron, figures which almost certainly underestimate the latter. By 1866 the concern had been extended by the purchase of the previously ill-managed interests of Richardson and Company at Shotley Bridge, consisting of a colliery, 27 puddling furnaces, and three plate mills (see Fig. 4.2).[4]

The 1864 reconstruction provided Consett with much more competitive conditions than ever before. The main reason was a change in financial circumstances. The new company acquired properties which had originally cost about £1 million; it paid only £295,318. Consett Iron's capital was only £400,000. Consequently, overheads for many years were low. The other main assets were neither material nor completely new, but they were now fully recognized for the first time. Reconstruction had brought in what the Newcastle correspondent of the *Colliery Guardian* characterized as 'spirit and intelligence' to a greater extent than ever before. Nor was the situation wholly unfavourable on the resources account. Though ore was expensive at Consett, the company had good coal which could be worked for the very low price of 2*s*. 0*d*. to 2*s*. 6*d*. a ton. The public responded enthusiastically to the opportunity to invest in the new limited company. Within less than 10 weeks of the first advertisement of the prospectus, all Consett Iron shares had been taken up. None could be purchased, even at a premium.[5]

The assembly and marketing situation of the works was being improved at the same time as the new company was taking shape. During summer 1864 work began on a railway from Consett to Blaydon, thereby giving direct access to the industrial areas of upper Tyneside. Through David Dale the company was strengthening its links with the new North-Eastern Railway Company.

[3] *MJ* 13 June 1862, p. 412; Tomlinson 1914: 562.
[4] *CG* 2 Aug. 1862, p. 94; Consett 1893.
[5] *MJ* 15 Jan. 1859, p. 49; *CG* 2 Apr. 1864, p. 259, 18 June 1864, p. 479.

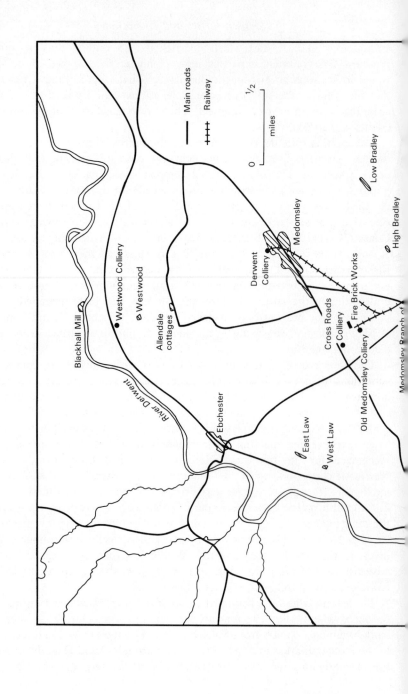

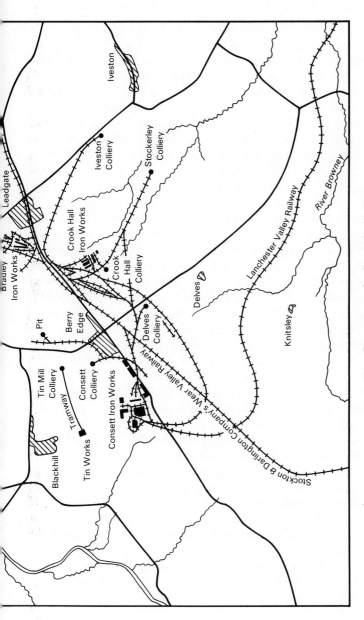

Fig. 4.2. The Consett area in the early 1860s

Already, having sold off some of its raw material properties, Consett had secured undertakings from others for long-term mineral supplies. This policy had been followed first as long ago as 1845, when the Stanhope area limestone interests were sold; twelve years later it had been repeated in the sale of the Upleatham mines to Pease and Partners.[6] By the late 1860s Consett was still using a mixture of iron ores to a greater extent than other north-eastern works. The company remained a major coal producer, by 1869 leading all the iron concerns in the region in the number of pits controlled, with 8 to Weardale Iron and Coal Company's 6 and the 5 owned by Bolckow and Vaughan.[7]

In spite of these favourable circumstances, as Teesside output grew, Consett Iron had to keep up a constant campaign of plant modernization, wage paring, and other means of cost reduction. Within two years from the first half of 1866 it managed to cut production costs per ton of plates from £7. 16s. 6.02d. to £7. 3s. 4.38d. Wage reductions represented 9s. 3d. of this decline.[8] Efforts to improve the cost structure included a reconstruction of ironmaking in the Consett area. By 1869 the number of blast furnaces there had been reduced from 17 to 6. This involved the closure of the establishments at Bradley and Crookhall and abandonment of the older furnaces at the Consett ironworks. Before the crisis of 1857, Bishopwearmouth works, though dependent on Consett for its pig iron, had almost as big a capacity for finished iron. In 1865 the Consett directors decided to concentrate their operations, and to that end they accepted an offer of no more than £13,000 from Henry Ritson for this Wearside works. It was their first retreat from tidewater.[9]

The Consett reconstruction succeeded. In contrast to the other coalfield concerns, Consett was able to record performances as good as those of Teesside ironworks. By early 1871, 97.6 per cent of the furnaces in the Cleveland district were active. In the rest of the North-East the proportion was only 86.3 per cent, but at Consett all six furnaces were operating.[10] The reputation for innovation and for quality of product was maintained. Here Whitwell's Patent Firebrick hot blast stoves were first installed, and as early as 1861 Charles Mark Palmer, already a major shipbuilder at Jarrow, had remarked that there was no better plate than that from the Consett mills. All this was reflected in the financial returns. From the formation of Consett Iron Company in 1864 to mid-1870, and after allowing for payment for the new blast furnaces, dividends averaged 10 per cent. They were even higher in the early 1870s. A few years later the writer J. S. Jeans was enthusing over the success of Consett, though in doing so he resorted to hyperbole which probably did less than justice to its earlier history: 'It would be impossible to

---

[6] Jeans 1875: 203; *CG* 6 June 1863, p. 448.
[7] Mineral Statistics of the United Kingdom; Transactions of the ISI 1869, p. 86.
[8] David Dale papers (NRRC 1027/3/31–2).
[9] CDM 29 July 1865.
[10] Mineral Statistics of the United Kingdom; *JISI* 1871.

find, in the whole industrial experience of this country, a greater contrast than that presented by the Consett Iron Works under their present and past management.'[11] In fact the successes of Consett in the 1860s were less than they might have been, and its high standing and the approbation which it attracted in the early seventies came only after a positive response had been made to a full and highly critical report on its operations which was undertaken in 1869. This efficiency survey was so important as to deserve separate and full consideration.

[11] *CG* 26 Mar. 1864, p. 241, 14 Sept. 1861, p. 170; Jeans 1875: 201.

5
Early Assessments for Development Planning

BY the mid-1870s Consett was the biggest producer of iron ship plate in Britain, able to turn out up to 1,300 tons per week, or more than three times the level of 14 years before. Within the region it had become an operation of major importance (see Table 5.1). Ironmaking plant had been reconstructed, six new 750-ton per week furnaces having replaced the 14 blast furnaces controlled in the 1860s, the largest of which had been capable of 340 tons of iron a week. Size and modernity of plant was still matched by quality and reputation.

Yet Consett was by no means always sure of itself. Its achievements had resulted from struggling with inherent difficulties and reacting to earlier failures in its own organization. Already (as on a number of occasions over the next three-quarters of a century) an assessment had been made of the prospects for future success on the old site. Consett's first economic consultant was Edward Williams.

Edward Williams (1816–86) was one of the numerous leaders of the iron trade derived from the great Dowlais Iron Works near Merthyr Tydfil. From 1842 to 1864 he was Assistant General Manager there, and after that their London agent. He was then recruited into the rapidly expanding iron trade of the North-East. Bolckow and Vaughan had been converted into a limited company, and as John Vaughan was no longer able to supervise their growing operations, Williams was engaged as General Manager. As head of one of their major regional competitors, he was commissioned in 1869 by Consett Iron to report on their operations. However, in a sense, Williams's contact with them and his evaluation of their prospects dates back to ten years earlier.

In February 1859, on behalf of Dowlais Iron, Edward Williams had visited works in the Midlands and the North-East. He subsequently wrote an account of this industrial tour (Dowlais 1859). At that time, Williams had been particularly impressed by the forges and mills at Charles Attwood's Tudhoe Works: 'these works are the finest I have ever seen.' He had not formed a particularly favourable impression of Bolckow and Vaughan operations and certainly not of the Walker Iron Works. Consett he described as a 'very fine works'. He concluded that generally the rails made in the north of England were far better than those which were being produced in Wales, and that Dowlais needed to look to its laurels to meet their challenge. Almost exactly

TABLE 5.1. *Pig iron production in Consett, County Durham, and the North Riding ('000 tons)*

	1848	1859	1865	1866	1867	1868	1869	1870	1871
Consett	29	85	87	92	62	78	68	87	100
Rest of Co. Durham	33	285	390	207	416	421	590	590	659
North Riding	—	260	486	546	641	699	766	917	1,030
TOTAL	62	630	963	845	1,119	1,198	1,424	1,594	1,789
Consett as %	46.8	13.5	9.0	10.9	5.5	6.5	4.8	5.4	5.6

TABLE 5.2. *Coal, coke, and iron production at Consett, 1868*

	Production ('000 tons)	Cost per ton		Transfer or sale price per ton		Profit per ton		Total profit
		s.	d.	s.	d.	s.	d.	£
Coal	492	3	4	4	2	0	10	20,509
Coke	179	n.a.		n.a.		1	6	13,452
Total profit on fuel division								33,961
Capital spending on new collieries and ovens								8,100
Net profit on fuel division								25,861
Net profit on fuel division less loss in ironworks[a]								23,460

[a] Loss in ironworks was £2,401.

Source: Williams 1869a.

ten years later he delivered his full consultant's report to the Consett directors.[1]

The Williams report opened up what were to be continuing issues of debate concerning a wise business strategy for Consett. It began by summarizing what Williams took to be the directors' reason for asking him to undertake the analysis—the problem of whether it would be more profitable to confine operations to production of coal and coke for sale, or to continue to consume the bulk of their fuel output in their furnaces, 'and, if the latter, to point out where, if at all, manufacturing costs are too high and expenditure of money or material might be saved' (see Table 5.2). His comparisons were usually based on the production and prices of 1867/8.

[1] Williams 1869a (NRRC 1027/6/7).

Williams discovered that slight losses were being made in ironmaking, amounting to about 1s. 0¾d. per ton of finished plates and rails rolled in 1867/8. Even so, he doubted whether they could sell all their coal and coke in the open market at prices comparable with those obtained for the smaller existing levels sold, for even for fuel sales they were not so well located as some of their competitors. The ironworks had made marginal losses, but 'the year's profit would have been even less than it was if your own iron works had not been your chief and best customer for coal and coke.' 'It seems therefore very desirable to keep on your ironworks if this can be done at a profit or even without loss.' He knew of no ironworks which could get cheaper coal, or obtain coke at less than 9s. a ton. Consett's cost was 9s. 7½d. On the iron ore account, not only were they at a disadvantage but they were not even making the best of a difficult situation. Cleveland ore, loaded on railway trucks, averaged 2s. 10d.; carriage to Consett added 4s. The ore was calcined there, by which time, not surprisingly, 'it has cost you very much more per ton than is paid by the average of the Middlesbrough ironworks.' By contrast, red ore (Cumberland haematite) could be had at a reasonable price, though 'you seem to be paying a high one at present.' Mill and forge cinder was another source of iron which they might use for their blast furnaces.

With costly ore, yet low-priced fuel, Consett should be able to make ordinary iron at 'moderate cost'. But Williams was critical of current practice. 'You are working with obsolete blast furnaces and stoves that waste fuel enormously.' He pointed out that their wealth of coal had caused them to be profligate in its use, above all in still operating small furnaces. Per ton of iron they were using over 26 cwt. coke and more than 10 cwt. coal, whereas 'proper' furnaces required 5 cwt. less coke, and by use of furnace gas could dispense with extra coal except for a little for calcining the ore. 'Here alone is a loss of about 4s. 4d. a ton on pigs, equal to about £13,000 on the 45,000 tons of finished iron made'; in short, if this improvement had been made, 1867/8 profits would have increased by over 40 per cent. He recommended installation of Whitwell stoves in place of 'the poor old stoves of which you have so many, wasting coal wholesale'.

In spite of his strong criticisms, Williams saw hope for a highly competitive future, partly because there were ample margins for economy in all departments. For instance, the forges and mills were being worked so wastefully that coal use per ton of puddled iron was 'certainly 25% too high'. By installing better boilers they could improve their general fuel efficiency. Moreover, having cheap coal, they were well placed to extend in further processing. In short, as Grunner and Lan had shown a few years earlier, though they could not rely on making haematite or Cleveland pig for sale, notwithstanding freights of 5s. a ton to ports, they should 'stand well' in finished iron. Indeed, 'Consett ought to make rails, plates and other wrought iron at less than the average costs in Yorkshire and Durham.' His report ended: 'In conclusion and after very careful consideration, the opinion I have come to is that Consett as a fuel producing property only is of far less value than as a complete apparatus for making wrought iron, and that fairly worked for the latter purposes it ought to yield a very good return for the moderate amount of capital your company has invested in it.'

Consett Iron acted promptly on some of these recommendations. The meeting which received the report, five days after it was dated in Middlesbrough, decided to build a third large blast furnace, calling on Whitwells's professional expertise. Within a further three weeks, the directors had agreeed to consider two more furnaces of this type. They also resolved to use in the furnaces as much as possible of their annual make of 50,000 tons of mill cinder. Consideration was even given to smelting with about 17 cwt. of cinder and 17 cwt. of haematite ore per ton of iron, for production of 1,000 tons of pig a week. The question of blast furnace burden was vital, for this determined whether the new furnaces should be of the dimensions then being employed on Teesside for reduction of Cleveland ore. One problem limiting their choice was that they were still bound by contract to take 50,000 tons of Cleveland ore a year or pay 6d. a ton fine on that amount. At this point the directors decided to ask Williams for further advice.[2] Again there was a sense of urgency. The second report was considered at a directors' meeting less than five weeks later.[3]

On this occasion Williams started from the perspective of the iron needs of their finishing operations. Not counting the newly acquired Shotley Bridge mills, these required 70,000 tons of pig iron a year. This he broke down in ways which pointed to the north-west coast as a continuing chief source of ore. He suggested that cinders should be costed at 1s. a ton, the cost of loading and moving. Their availability kept the cost of iron down, but used alone, they were an unsuitable material. For very strong pig iron he recommended that haematite should be unmixed with other materials. Costs he estimated as shown in Tables 5.3, 5.4, and 5.5.

With the better coke rates provided for in his first report and taking into account other improvements, costs for pig iron could be cut to about 41s., or no more than 80 per cent of the current level. Such a figure pointed to a highly competitive situation, for it was several shillings a ton below the price paid by those Cleveland wrought iron firms which bought their pig iron in the open market and, as it must have pained him to have to admit, 'even more below the cost of the pigs at the most favoured of the forges and mills in Wales'.

He recognized that the proposed furnace charge was different from that used in Cleveland, and that therefore Consett should not build furnaces as big as Cleveland's. Calcining of ore at the mine in Cleveland could have produced further savings by reducing the volume of material shipped, but he did not believe the landowners would allow this. He repeated the advice of his first report on the replacement of the old hot blast stoves, and emphasized the desirability of tackling improvements in the forges and mills as well as in

[2] CDM 3 Apr. 1869.   [3] Williams 1869b (NRRC 1027/6/7).

TABLE 5.3. *Consett capacity for balanced operations, 1869*

	Tonnage	Conversion Rate
Pig iron	70,000 }	Pig to bar, 22.5 cwt. per ton
Puddled bar	62,000	
Rails and plate	50,000	Bar to rails and plate, 24.8 cwt. per ton

Source: Williams 1869b.

TABLE 5.4. *Iron burden at Consett for production of 1,350 tons of pig iron weekly, 1869*

	Tonnage	Iron content (tons)	Cost per ton		Total cost (£)
			s.	d.	
Mill forge cinder	600	300	1	0	30
Cleveland ore	1,000	300	6	10	342
Haematite ore	1,250	750	20	0	1,250
TOTAL					1,622[a]

[a] Equal to 24s. per ton of pig.

Source: Williams 1869b.

TABLE 5.5. *Ironmaking costs at Consett, 1869* (per ton of pig iron)

	s.	d.
Actual costs (approx.)	51	0
Attainable costs:		
Iron content of burden	24	0
Coke at 20 cwt. per ton	9	8
Limestone	1	0
Other charges	6	4
TOTAL	41	0

Source: Williams 1869b.

ironmaking. The large savings which might result were outlined. Current practice was clearly very poor, in spite of Consett's attractive dividends and successful appearance to the outside world. The puddling forges were 'the weakest part', worked by a needlessly costly system. In them 31 cwt. of coal were used for each ton of puddled bar, 'fully half a ton too much', labour costs were 'fully' a third above what ought to be the level, and fettling at 18s. 6d. was 'about three times what it ought to cost'. In the rail mill, hammering should be replaced by rolling. Williams ended his second report, 'it is with much regret I feel compelled to express so unfavourable an opinion of your forge and mill operations and I only do so because I could not with honour withhold it.'

Two weeks later the directors met again to discuss Williams's views.

TABLE 5.6. *Estimated proportions of pig iron sold and used in making finished iron at Consett and in the North-East, 1859, 1866, 1871/2*

	% of total output	
	Sold	Used in puddling
Consett:		
1859	51	49
1866	43	57
1871	22	78
North-East:		
1872	51	49

Jonathan Priestman, as General Manager, responded, apparently aggressively, to the charges about inefficiency. He pointed out that Bolckow's experience was not necessarily applicable to Consett, for, whereas plate was to them twice as important as rails, in the Teesside firm, rails were much more important. (In September 1867, the tonnage of rails at Witton Park was twelve times as great as that of plates.) As for some time past, Consett was using more of its pig in further finishing operations than the ironworks of the North-East generally (see Table 5.6). Priestman then turned to furnace operations. He pointed out that Witton Park used mainly Cleveland ore, but also some red ore, and yet had iron costs on the ore account well in excess of those which Williams predicted for Consett. However, Priestman's own figures seemed to show that Consett's pig costs were at this time as much as 25 per cent higher than Williams had suggested. His counter-attack was not successful, and within six weeks of his fierce response to the arguments of the two reports, Priestman had resigned. The Priestman family went on to invest in coal-mining in the district north of the lower Derwent in County Durham. There, in Chopwell, their interests and those of Consett Iron were to come together again 20 years later. More immediately, Priestman's replacement constituted another decisive step towards Consett's future success. In this too the Board showed an impressive sense of urgency.

On 3 July 1869, the Consett directors decided to advertise for a new General Manager. The notice they agreed read:

Ironworks General Manager wanted. Wanted by the Consett Iron Company Ltd., a Gentleman thoroughly competent to undertake the Practical as well as the Commercial Management of their extensive Ironworks (blast furnaces, rolling mills, etc., etc.,) in the county of Durham. Written applications may be forwarded up to July 22nd under cover addressed to David Dale Esq., Darlington. They will be received in confidence, and must state fully the applicant's experience, qualifications and references.—To a first-class man a liberal salary will be given.[4]

[4] CDM 3 July 1869.

Sixty-four applications were received, but were quickly reduced to a short list of six. By 18 August, it had been decided to appoint William Jenkins. He was to be paid £1,500 a year plus a commission of up to £500 and was to occupy Consett Hall free of charge for rent, coal, or gas.[5]

Jenkins was another product of Dowlais. He was said to have spent 34 years working his way through most departments there—which if correct meant that he had started work at the age of 10. His advancement was described as by 'the old healthy means of steady industry and efficiency'. After the death of Sir John Guest in 1852, trustees had run Dowlais. The key figures had been Edward Williams on the technical side, and, in the overall control of the commercial side, William Jenkins. It must be assumed that Williams's advice was important in Jenkins's appointment to Consett, just as their joint transfer to the North-East was eloquent testimony to the shift of competitive advantage against the great ironworks of South Wales.[6]

Before he moved Jenkins was warned by some outsiders that Consett had no long-term future. It was suggested to him that, notwithstanding Williams's report, iron could not be produced successfully there and that the company should confine itself to coal production.[7] However, having visited the plant and studied its raw material situation and costs, Jenkins accepted the challenge. Initially, and within the confidences of the company itself, he seems to have been cautious about prospects, recognizing that the struggle would not be an easy one. Some of these thoughts were revealed just before Christmas 1869. The occasion was a meeting held in the Lecture Room attached to the Consett Co-operative Store to present a farewell address to Jonathan Priestman. Speaking as his successor, Jenkins expressed a wish that agents and workmen would do what they could for the Consett Company first and themselves afterwards, for this was essential for their survival.

The first thing to look to was to keep capital entire, for without that they could not be prosperous. If that was safe there was some hope of their remaining on the spot to which they were attached; if not, they might soon be scattered to the winds. So far as he was able, he would ally himself with them closely in anything that could promote their interests.[8]

In the quarter-century in which he was to manage Consett, Jenkins played a key role in transforming it into one of the most successful iron and steel firms in Britain.

[5] CDM 28 July and 18 Aug. 1869.
[6] ICTR 12 Jan. 1894, p. 46.
[7] CG 5 May 1882, p. 707; Neasham 1882.
[8] CG 24 Dec. 1869, p. 612.

6

Re-equipment and the Reorganization of the Supply and Marketing Situation to the mid-1870s

IN his earliest years with Consett Iron, William Jenkins was involved in a major modernization process which covered the plant and its sources and modes of raw material supply. At the same time output levels increased, though the break with the previous era was by no means sharply marked in this respect (see Table 6.1).

Jenkins reported regularly to his directors on progress in the installation of new plant. The old Consett furnaces had gone by 1870, Bradley works were dismantled, and the Crookhall iron plant was soon abandoned. Two years later he was stressing the need for another large furnace at Consett, the sixth of the new units to be built there. He also followed up Williams's suggestion and installed more Whitwell stoves. By 1873 the company had seven rolling mills, four of them rolling plate, though rail output was of broadly similar size.[1] Markets for plate were largely, though far from exclusively, within the region; the outlets for rails were mainly overseas. In both products the company enjoyed a high reputation at a time when iron rails in particular were often a cause of considerable dissatisfaction to their buyers. For instance, in August 1870, Jenkins was able to report: 'Touching the question of quality of Finished Iron made, I may remark that during the past year the rails made for Russia have been accepted without serious inconvenience and those for America are well, and I may say highly spoken of by the Purchasers and Inspectors concerned in the reception of them.'[2]

Big developments were begun in fuel supply, following a period in which expansion in production of coal and coke had already been much more pronounced than in metal. In 1865 they had owned only 500 of the 7,004 coke ovens in County Durham and were reckoned to make only a little over 7 per cent of its coke. This was supplied not only to the Consett furnaces but to those on the North-West Coast and to the railway companies for locomotive use (Hall 1865). In the early seventies new ovens were built, the waste heat

[1] CDM *passim*; *JISI* 1870, p. 142; CDM 5 Mar. 1872; *PCE* 98 (1889), 407.
[2] David Dale papers, 5 Aug. 1870 (NRRC 1027/3/31–2).

TABLE 6.1. *Output of main products by Consett Iron Company, 1865–1871* ('000 tons)

	1865	1866	1867	1868	1869	1870	1871
Coal	354	394	413	492	496	496	523
Coke	132	144	140	179	189	191	207
Pig iron	87	92	62	78	68	87	100
Puddled iron	48	48	41	60	59	74	87
Finished iron:							
Plate	18	18	19	30	25	28	34
Rails	14	19	12	14	21	30	32

Source: David Dale papers (NRRC 1027/3/31–2).

from some of them being utilized for brickmaking. By 1877, the locomotive market had been lost as coal-firing took over, but coke was being delivered to additional markets on Teesside and to foundries in the Midlands. The annual output capacity for coke was now 350,000 tons. This expansion required a comparable growth in coal production. After buying Milkwell Burn colliery near Lanchester, Consett Iron in 1873 leased from the Earl of Durham an area further down the Browney valley, where over the next two years they sank Langley Park pit. Beehive ovens to coke all the coal raised were installed the following year. By 1877 Consett had ten collieries capable of raising one million tons of coal a year.[3]

Iron ore supply, being outside the company's direct control, proved more difficult to expand or to streamline. Consumption of Cleveland ore had been falling rapidly in the second half of the sixties, and by 1868 the tonnage from Furness and from Cumberland was five times as great, and in terms of iron content much more important still (see Tables 6.2 and 6.3). Attempts were made to improve arrangements for the Cleveland ore still obtained from Upleatham. This had been calcined at Consett, which involved considerable waste in carriage.[4] In 1870 Dale and Jenkins looked into the possibility of calcining before shipment, but, as Williams had suspected, this proved impracticable. Lord Zetland only permitted calcining at the mine on terms which would so increase costs there as to cancel out any gain on the carriage account. If they chose to build an independent calcining plant in the Cleveland area away from Zetland's properties this would involve both ground rent and increased freight charges due to the break of bulk which would result. Calcining, it was concluded, would have to remain at Consett.[5]

---

[3] Mountford 1974: 7; *MJ* 10 Mar. 1887, p. 260; *CG* 23 Mar. 1877, pp. 456, 457; *Engin.* 21 Sept. 1877.

[4] Calcining is the roasting of ore to drive off moisture and carbon dioxide, which reduces it to ferric oxide. The saving in tonnage moved may be appreciated from Pattinson's mid-century analysis of Cleveland ore as containing 4.4% water and 22% carbon dioxide.

[5] Wright 1880; CDM 3 Sept. 1870.

Re-equipment and Reorganization

TABLE 6.2. *Consumption of Cleveland ore at Consett, 1864–1866 and 1868*

Year	Consumption ('000 tons)
1864	119.9
1865	64.4
1866	46.6
1868	below 15.0

Source: Consett Records and Mineral Statistics of the United Kingdom (annual).

TABLE 6.3. *Deliveries of iron ore to Consett by the North-Eastern Railway, 1868*

Place of origin	Quantity of ore delivered ('000 tons)
Cleveland via Stockton and Darlington	13.8
Newcastle	1.1
Ulverston	5.7
Whitehaven	70.2
TOTAL	90.8

Source: Mineral Statistics of the United Kingdom (annual).

Arrangements with the North-Eastern Railway were a continuing problem, but were also seen to present opportunities for cost-cutting. David Dale was chairman of NER's Darlington section in 1871. The connection was severed the following year, but he was an NER director again by 1881. Consett knew how important its business was to the railway, and always proved willing to exercise the leverage which this gave it. For example, in spring 1869, complaining of freight charges for both raw materials and on finished products, the directors wrote to the NER,

It must be recognised that all our traffic passes over the North-Eastern system and the greater part of it of necessity for very long distances and whatever promotes our trade promotes the traffic of the North Eastern company, no part of this advantage being diverted to other companies or absorbed by water carriage. We may add that but for the operation of the rates set forth in this letter we should have spent a considerable sum of money in opening up our coal royalties and extending our operations.[6]

It was already a well-worn approach to the demand for lower charges, but was often to be repeated.

[6] CDM 8 May 1869.

TABLE 6.4. *Reductions in freight charges on undamageable iron from Consett, granted by the North-Eastern Railway in 1865*

Destination	New rate as % of old
Blaydon	60.0
Elswick	70.0
Gateshead and Newcastle (for local use only)	85.7
Darlington, Stockton, Hartlepool, & Middlesbrough	83.3
Clydeside	90.9
Hull (for local use)	96.7
—— (for export)	91.6
Merseyside (plate)	84.2
—— (other iron)	89.4

Source: CDM 2 Sept. 1865.

As early as the month when the Consett Iron Company was formed, Dale and one of his fellow directors approached the North-Eastern Railway with a request for a general rate reduction. They were rewarded with partial success eighteen months later, with some substantial cuts in rates on finished iron, for delivery both within the region, and to more distant markets, delivery to which also involved movement over other railway systems. To Elswick the old rate was 5s. a ton, the new rate 3s. 6d.; to the Mersey estuary the charge for carriage of plate iron fell from 15s. 10d. to 13s. 4d. (see Table 6.4).

The NER refused at this time to reduce the rate on Cumberland iron ore, then being used to the extent of 15 or 16 cwt. per ton of pig, and deferred a decision on the freight charge for Cleveland stone.[7] There were soon further negotiations on finished products, on coal, and on both Cumbrian and Cleveland ore.

Even after the substantial concessions of 1865 Consett Iron was not satisfied with the charges for deliveries to its customers. In April 1869 they addressed the North-Eastern again. First they raised the matter of the rate to Tyne Dock, in comparison to charges to tidewater for their rivals. For Consett the charge for the 23-mile journey, and for dock dues, averaged 5s. a ton for the various categories of finished iron. By contrast, 'the position of our competitors in Middlesbrough enables them to put on board at the cost of a few pence.' Given their relative locations such differences do not seem surprising, but a more forceful argument could be advanced about the fact that in selling to other districts of Britain, too, they were at a disadvantage in competition with the same firms. In deliveries to Liverpool this differential amounted to 10d. a ton, and in those to Hull, to as much as 1s. 8d.[8]

They believed that they had cause for complaint too over coal and coke. It

---

[7] CDM 2 Sept. 1865; 6 Jan. 1866.
[8] CDM 8 May 1869.

was true that the railway company had provided them with physical accommodation for their expanding output—in 1874, for instance, doubling the line up the Browney valley to the Langley Park sinking—but the rates charged were high. In 1869 a Consett submission identified their difficulties in supplying Middlesbrough as compared with coal producers in the Darlington section of the field.

They maintained that Consett should have a lower ton-mile rate than pits in the Crook area, because of the longer haul. (However, rather illogically, at the same time they were complaining that whereas Crook and Auckland area producers were paying 0.75d. per ton mile on their 24-mile haul to Teesside, they were being charged 1.5d. per ton mile to Elswick, 14 miles away.) On coal a zonal rate of 1s. 9.20d. per ton applied as far north as Tow Law; lying beyond this Consett was at a disadvantage in deliveries to Teesside. Even in shipments of coal to the Tyne their rate was only marginally less than the Tow Law rate to the Tees, though the distance involved was less than two-thirds as far. For blast furnace coke Tow Law to Teesside shipments were 8d. cheaper than Consett to Tyne shipments, even though the length of the haul was 60 per cent greater.[9] In these instances Consett seemed to have legitimate cause for complaint.

The NER allowed some concessions on Cleveland ore in 1866, at which time Consett gave notice that it wanted to move from the use of chaldron wagons, with their small capacity, to trucks. In April that year the freight charge for the 53 miles from Marsh House Junction, the point to which Peases then delivered the ore for Consett, was cut from 4s. to 3s. 10d. a ton. Even the new rate was a little short of 0.87d. a ton mile, whereas Consett maintained that no charge over 0.75d. was reasonable.[10] A few months later they won more concessions. Pease's ore having been placed in trucks at 2s. 10d. a ton was now to be carried all the way from either Upleatham Bank Foot sidings (56 miles) or from Hob House Junction (57 miles) for 4s. a ton. They were still not satisfied, complaining a little over two years later that for journeys of such length the rate should be still lower, for a high charge 'imposes a heavy restriction upon the first branch of manufacture'. In 1870 they pressed for the rate on Upleatham stone to be cut to 3s. 8d.[11]

A long battle was also fought over the north-western haematites, with which Consett had a variety of problems. In 1868 the tonnage of Ulverston and Whitehaven ore delivered to Consett was five times as great as that of Cleveland ore. In terms of iron content the relative contribution of the North-West was even more predominant. The 1869 freight rate for the 117 miles from Cleator Moor (72 miles of which were over North-Eastern tracks) was 18s. 6d. a ton. A little later, in the hope of cutting some of this cost, they

[9] CDM 8 May 1869.
[10] CDM 3 Feb. 1866, 6 Jan. 1866, 7 Apr. 1866.
[11] CDM 29 Jan. 1867, 8 May 1869, 1 Mar. 1870.

contemplated, but in the end did not build, what was essentially a revival of a link planned nineteen years before, a short length of railway from Carrs Colliery to join the Carlisle to Newcastle line at Stocksfield.[12]

Consett Iron continued to be interested in supplies of home ore. In 1873 they contemplated the purchase of Liverton mine in Cleveland, but decided against it. Even so, next year they paid the pioneer of the Cleveland ore trade, John Marley, £210 for a report on Liverton, Kilton, and other mines in that district. This was a time of high activity in the iron trade, and Consett had contracts to take 26,000 tons from Stevenson, Jaques and Co., paying 4s. 9d. for it at the mine.[13] In the same year they took a momentous step in another direction, choosing to cut through the complexities and difficulties of home ore supplies by organizing the delivery of higher-grade ores from overseas.

Two other concerns with home ore problems, the Dowlais Iron Co. and Krupp of Essen, joined with Consett Iron and a Spanish concern, the Ybarra Co., in 1873 to form the Orconera Iron Ore Company. Their purpose was to work the haematites of the Bilbao district. Though in the mid-seventies the Carlist wars disturbed the course of development for a time, the project went ahead. Not only was the ore rich and pure, but it was worked by lowly paid labour: 'the Spanish work from sunrise to sunset' was how a rival project in the same area was described.[14] Large-scale shipments began from shipping staithes specially designed to load big steamers quickly. The whole scheme was a prototype for the long-distance sea transport of iron ore; Consett Iron was again pioneering. As with its employment of Cumbrian ores, this use of high-grade Spanish iron ore—in which it was to retain an interest for 75 years—helped Consett to produce better iron than could be made from Cleveland ore. (Bolckows also acquired Spanish ores at this time, and other Teesside firms were soon buying them for use in their furnaces.)

In the first week of January 1881, just over 11 years since he had first spoken of his hopes for Consett, Jenkins provided an assessment of their situation. The occasion was a meeting and dinner over which he presided of managers, heads of department, and foremen, held in the Consett National School. He traced the progress of Consett's modernization and extensions. The old, inefficient furnaces had been swept away, so that 'they had now, he was proud to say, as fine a blast furnace plant in every respect as any in this country', and indeed in the year just finished the Consett No. 1 furnace had recorded the highest weekly average output of any furnace in England. When he had first arrived their output of finished iron plates and rails had been barely 900 tons a week. It had reached 1,700 tons in the mid-seventies and since then they had

[12] CDM 8 May 1869, 5 Oct. 1875, 3 July 1869.
[13] CDM 2 Dec. 1873, 9 June 1874, 6 Oct. 1874.
[14] CG 26 May 1876, p. 832.

further increased it.[15] However, in the last few years the iron rail business had collapsed. This not only caused them to pay more attention to plate, but was a factor which was edging them into steelmaking. That development was eased by their possession of rich, low phosphorus ores from Spain.

[15] *MJ* 7 Jan. 1881, p. 21.

7

The Great Crisis: Collapse in the Finished Iron Trade and Movement into Steel Production

During the third quarter of the nineteenth century Consett Iron made great strides in finishing the pig iron which it produced. By 1863, with 99 puddling furnaces, it was already well ahead of every works in the region, though together Bolckow and Vaughan's Witton Park and Middlesbrough works contained 139 furnaces. Consett had 171 puddling furnaces by 1871, by which time its rolling mills were capable of 800 tons of rails and 1,300 tons of plate a week.[1] Coal production was steadier than that of iron and to that extent was a valuable bell-wether, but the larger share of profits was in iron (see Table 7.1).

Until the mid-1870s steelmaking seemed to pose no more than a distant threat to the prosperity of the north-eastern iron firms: at that time it became a major factor in the terrible depression which threatened to sweep them out of existence. Henry Bessemer had announced his process of steelmaking to a largely sceptical metallurgical world. Early experiments with it in existing ironworks and forges were almost all disasters. One early North-East Coast experiment was reported in the September 1857 issue of the *Gateshead Observer*: 'part of an ingot of iron, made by the Bessemer process, has this week been tried at a forge on the Tyne. It was placed, red hot, under Nasmyth's hammer and beaten. Instantly it flew in all directions ....' (Birmingham 1857). This beginning, though so inauspicious, was to prove the distant harbinger of a revolution. Many of the existing works did not survive that revolution; Consett made the transition, not without some disturbance, but in the end successfully.

Theoretically any movement to steel was to the disadvantage of an interior, coalfield location. The puddling forge used large amounts of coal; steelmaking, especially in Bessemer's process, but to a lesser extent also in the Siemens open hearth process, was more fuel efficient. In the palmy days of the finished iron trade, coal use per ton of product was about 3 tons. By the end of the century 12 cwt. or so of coal was required for every ton of steel ingots, and probably 15 cwt. would represent the average for the manufacture of plates or rails from

---

[1] *Engin.* 7 Feb. 1873, p. 102; Louis 1916: 59.

The Great Crisis

TABLE 7.1. *Profits in Consett operations, 1870*

Source of profit	Profit (£)
Pig iron and castings, puddled bar, plate, and rails	72,114
Coal	17,692
Coke	10,643
House rents	3,853
Fireclay, royalties, land rents, etc.	2,417
Total net profit	106,719

Source: Whellan 1894: 1231.

the pig iron stage onwards.[2] (At the same time, fuel economy was still continuing in blast furnace practice, though the results were less spectacular than this contrast between the rate in finished iron and in steel. A representative of Bell Brothers noted that in 1857 their coke consumption per ton of iron made at Port Clarence was 32.4 cwt. By 1860, with no changes of the furnaces or stoves, they had already cut the rate to 27.8 cwt., and by the mid-1860s to 26.9 cwt. Although performance varied by works, the average consumption of coke per ton of iron in the Cleveland district fell from 38 cwt. in 1860 to 25 cwt. by 1879.[3]) Such changes tended to help locations ill favoured for access to coal. However, date and scale of entry, and still more, efficiency of operation, proved more important factors in the new economic geography of the industry which now centred on steelmaking. As always, changes in locational influences were permissive rather than deterministic; they provided challenges and opportunities, not cut-and-dried consequences. Consett Iron Company proved able to master the new circumstances by taking initiatives. Its pioneering resulted in major commercial rewards.

The early failures of the Bessemer process were in the main due to use of unsuitable pig iron. When this problem had been tackled successfully through the purchase of low phosphorus irons, steel began to find uses in fields formerly monopolized by wrought iron. As early as 1858, at Dowlais, William Menelaus and Edward Williams had rolled the first Bessemer rails. Within three years the Lancashire engineer Daniel Adamson and the Oldham textile engineering firm of Mather and Platt were employing steel boiler plate. (Steel ship plate was then used only for special purposes.) In 1861 Charles Attwood installed three small Bessemer converters at Tudhoe, only 15 miles from Consett. There he produced some rails in the new material, and also rolled steel plates for Henry Bessemer. However, as late as 1873 Tudhoe was the only steel plant of any significance in the North-East; at that time there were 22 finished ironworks there. Consett's location had meant that it was well placed to provide the fuel needs of the puddling forge. Use there of large tonnages of

---

[2] *Royal Commission on Coal Supplies* 1904: 294.
[3] Bell 1883; *CG* 3 Dec. 1880.

west-coast haematite favoured it for the production of Bessemer grade pig iron when the highly phosphoric nature of Cleveland ore prohibited its use for that purpose. Establishment of the Orconera Iron Ore Company gave Consett direct control over its supplies of haematite. More positive incentive in the direction of steel manufacture came from advances in steel by rival concerns and, most decisive of all, from the collapse of the market for finished iron.

In the late 1860s a number of British firms, especially in Sheffield, were doing a rapidly rising business in steel rails and to a lesser extent in steel plate. In 1871 Bolckow Vaughan took the first step in the same direction when they bought a steel plant at Gorton, Manchester for large-scale experiments with the Bessemer process, with the expectation that this would lead on to construction of steelworks at Eston. The following year they joined the leading Sheffield steelmaker, John Brown, in acquiring interests in Bilbao ore. Consett was aware of, and already to some extent responding to, the new circumstances. As early as September 1872 Jenkins reported that they had made 2,000 tons of Bessemer iron, but that sales had proved difficult. Two and a half years later the directors took an important decision: 'Some consideration was given to the advisability of adding to the Company's business by the erection of steelworks and it is resolved that Mr Jenkins be encouraged to prosecute his enquiries on the subject and report to the Board.'[4] By this time the finished iron trade both in the region and in Britain as a whole was in a state of crisis, a situation which for a major concern like Consett Iron was soon seen as making entry into steel manufacture a necessity. The crisis was the product of combined trade depression and an accelerating switch of customers from iron to steel.

For most purposes steel was recognized as a material superior to iron; the critical factor in their respective sales prospects was price. Progressively through the 1870s the price differentials which had limited the use of steel were pared away. In response to this decline in the price for a better product, the demand for iron rails was the first to collapse (see Table 7.2). By 1876 Britain sold 174,000 tons of steel rails overseas; only 16,000 tons less than sales of iron rails. Over the next two years steel rail prices fell by a further 27.5 per cent and their differential over iron rails narrowed from £1.14s. 2d. to 10s. 5d. a ton. The regional effects were devastating. The Cleveland district made 265,000 tons of iron rails in 1874; three years later output was 37,000 tons.[5] Many firms failed. Some of them were reconstructed, but others did not survive. In a number of cases a switch was made from rails to iron plate, in which the competition from steel was as yet much less severe. Everywhere there was depression of business, idle plant, and unemployment. By spring 1879 the great railway engineer Thomas Brassey was writing what was in effect an obituary of the wrought iron rail.[6] 'The discoveries of science are

---

[4] CDM 28 Sept. 1872, 2 Mar. 1875.
[5] BITA 1878, pp. 27, 37.
[6] *The Times*, 8 Jan. 1876, p. 6; *the Nineteenth Century*, May 1879, p. 795.

TABLE 7.2. Prices of Bessemer steel and of wrought iron rails and plate, 1873–1881

	1873			1876			1878			1880/1		
	£	s.	d.	£	s.	d.	£	s.	d.	£	s.	d.
Rails:												
Wrought iron	11	4	4	6	0	10	5	2	1	5	6	2
Steel	15	7	6	7	15	0	5	12	6	7	0	0
Plate:												
Wrought iron	12	10	0	7	3	5	6	2	6	6	1	8
Steel	21	0	0	18	10	0	12	8	1	11	14	5

Note: Steel prices are averages for year; wrought iron prices are averages for last quarter of year. In the last column, steel prices are for 1880, wrought iron prices are for 1881.

Source: British Iron Trade Association Reports.

frequently attended with ruinous consequences to large masses of workmen. The Bessemer process has led to an almost complete discontinuance of the use of iron rails.' His estimate was that of 45,000 puddlers formerly employed in the whole of the North of England and in Wales, less than half were still at work. A few weeks later the state of the north-eastern iron firms and their workers was surveyed by the *Colliery Guardian*.[7] It reckoned that 7,000 fewer men were employed than five years before, without making allowance for the additional losses in jobs in iron ore mining or in coal or limestone working. Recognizing that a crisis existed, it analysed its economic and social implications in an interesting fashion. In the course of its remarks the journal wrote:

In part the surplus labour has been taken up by some other trades. The large steelworks at Eston has been put into active operation in the interval, though the amount of labour needed in the steel manufacture is small compared to that in the manufacture of iron; yet there has been a large number of workmen there employed; whilst the new foundries at Port Clarence, Stockton and in other places, as well as the brisk shipyards, have found employment for some of the otherwise idle, unskilled labour. There has also been an exodus of labour from the district.

All could see that for the men and for their employers alike the situation was of an unprecedented seriousness. By late summer 1879 only 38.8 per cent of the puddling furnaces which had existed in the district were still at work. Of this overall figure of 2,158, 838 were working, 432 were standing, 821 were in plants which had failed and had not been restarted, and the remaining 67 had already been pulled down in order to be replaced by steelworks.

At the beginning of 1876 Consett was said to be the only plant in the North-East working full-time. Over three and a half years later it was still in that exceptional position. At what was now the only bigger finished ironworks in the region, that of the Darlington Iron Company, just over one-fifth of the

[7] *CG* 4 July 1879.

furnaces were active.[8] Consett had extended its finished iron capacity from 900 tons a week in 1869 to 1,700 tons by the mid-seventies. It was helped in the difficult trading conditions of the later parts of that decade by the fact that it was more than usually significant in plate. By 1877 for instance it had seven plate mills and only one mill on rails, though the disparity in tonnage was much smaller: 1,200 tons to 800 tons weekly. Notwithstanding its product mix and its high operating rate, it suffered severely from the depression. In the year 1872/3, when trade was buoyant, Consett profits were £305,000. Four years later—though Consett was fortunate to be still profitable at all—the total was £83,000, and by 1878/9 it was below £56,000.[9] For a time there were rumours that the company was contemplating a large-scale involvement in steel rail manufacture as a way out of its difficulties. However, the way of advance which was chosen appeared for a time to be much less spectacular than that. At the end of the seventies, Consett Iron along with other firms seemed to have found a secure refuge in the production of iron ship plate. The single rail mill was converted to roll plate. By 1878 Consett was already being described as 'the largest plate making works in the world'. In 1880 it was sending out 1,600 tons of ship plate a week. The following year weekly shipments were 2,000 tons. Such achievements were helped by the organizational links which the company had with north-eastern shipyards. (David Dale was for a time Vice-Chairman of the first big shipbuilding syndicate in the North-East.) Even so, plant efficiency was a major factor in commercial success. Together, product specialization and competitiveness brought further success going beyond that of other favoured concerns. In 1879/80, while Palmers, both a major producer and consumer of iron, felt able to declare a dividend of only 4 per cent, Consett managed 20 per cent.[10] Profits rose from the £56,000 of 1878/9 to £128,000 three years later and £130,000 in 1882/3.

After this the situation worsened. By 1884/5 Consett Iron profits were again as little as £60,000. The reasons were in some respects very similar to those of a decade before, in this instance however affecting the plate business rather than rails. There was a general depression of demand and an advance of the competing material made from steel. The gross tonnage of merchant shipping built in the United Kingdom rose by 70 per cent from 1880 to 1882. It increased by another 11 per cent, or 100,000 tons, in 1883. During the next three years there was a decline, and the tonnage built in 1886 was only 71 per cent of the total of six years before. In 1880 over 89 per cent of the merchant tonnage built was of iron; in 1886, only 46.4 per cent (see Table 7.3). In 1880 Consett was still delivering large tonnages of iron shipbuilding material to yards on the Clyde. In the 12 months to June 1881 this business amounted to 31,041 tons. However, at the same time some steel plates were being brought

---

[8] *JISI* 1879, pp. 544, 545.
[9] *CG* 5 Sept. 1873, p. 183; *The Times*, 21 Aug. 1877, p. 5.
[10] *The Times*, 9 Sept. 1880, p. 4.

TABLE 7.3. *Tonnage of merchant vessels built in and added to the register of the United Kingdom, 1875–1890*

Year	Total tonnage (gross tons)	Built of iron	Built of steel
1875	542,300	487,500	—
1880	545,506	487,404	38,164
1883	1,027,937	856,990	155,745
1886	387,208	179,914	191,561
1890	968,469	46,055	913,384

Source: Lloyd's Registry, quoted P. Watts 1902.

into North-East Coast yards from steelworks in the west of Scotland and South Wales. At this time the price for these plates was still double that of iron plates.[11] By 1882 the North-East had become an anomaly, for here competition between iron and steel in plate markets was not so prominent as elsewhere and iron plate production still exceeded that in steel. Yet for producers like Consett, margins had begun to narrow, as rising prices for inputs, including wages, pushed up overall costs for iron plates, while at the same time high levels of production in relation to demand kept prices down.[12]

Already during the expansive days for iron plate, Consett management had recognized the way in which demand was likely to change and had begun to provide for it. In 1880 the *Colliery Guardian* acknowledged that the North-East needed to foster the expansion of its capacity to consume more of the iron which it made and ought to respond more positively to the opportunities of the steel age. Consett was seen to be already leading the way. 'The Consett Iron Company—probably the largest producers of plate in Britain—is the one firm in the North which seems to be taking steps to meet the demand . . . .'[13] That year Consett Iron decided to spend £10,000 on experimental steel works, and in 1881 Jenkins made visits to the most up-to-date steel plants in the country. In 1881/2 Consett Iron profits were 26 per cent, but the decision was taken to go ahead with steelmaking on a commercial scale. Two 13-ton open hearth furnaces were given their test run in June 1883. By this time there was a good deal of leeway to make up, both within the region and even more in comparison with other areas of Britain. In the year to June 1882, two Scottish works, Hallside and Blochairn, had made 91,000 and 61,000 tons of steel respectively. Not until 1885 did the North-East as a whole equal their combined output. Jones Brothers of the Ayrton Works in Middlesbrough accepted orders for steel plate in 1882, but they were relatively small producers and had to buy their steel, much of it from the Darlington Iron and Steel

---

[11] *MJ* 19 Jan. 1878; *CG* 1 Oct. 1880; *I* 4 Jan. 1884, p. 6; Consett 1893: 32; *Engin.* 4 May 1906, p. 433; David Dale papers (NRRC 1027/3/31–2); *CG* 12 Nov. 1880, p. 771.
[12] *CG* 24 Nov. 1882.
[13] *CG* 9 July 1880, p. 66.

TABLE 7.4. *Consett Iron Company profits, 1870–1887* (financial year to 30 June)

Year	Profit (£ '000)	Year	Profit (£ '000)	Year	Profit (£ '000)
1870	37	1876	86	1882	128
1871	102	1877	83	1883	130
1872	195	1878	58	1884	86
1873	302	1879	56	1885	60
1874	304	1880	104	1886	72
1875	215	1881	195	1887	96

Sources: *MJ* 11 Aug. 1883, p. 927, 6 Aug. 1887, p. 962; Whellan 1894: 1231.

TABLE 7.5. *United Kingdom and North-East Coast open hearth production, 1884–1890*

Year	Production ('000 tons)	
	United Kingdom	North-East
1884	475	16
1885	584	75
1886	694	124
1889	1,429	437
1890	1,564	470

Sources: *CG* 21 Aug. 1891, and BISF.

TABLE 7.6. *The condition of North-Eastern ironworks and blast furnaces, 1892*

County	Ironworks		Furnaces	
	Total no.	In operation	Total no.	In operation[a]
Northumberland	2	1	5	1
Durham (interior)	10	4	32	$9\frac{3}{12}$
of which Consett	—	—	7	$5\frac{3}{12}$
Durham (coastal)	4	3	22	12
North Riding	19	18	85	$50\frac{8}{12}$

[a] Fractions indicate operations for portions of the year.

Source: *Mineral Statistics of the United Kingdom.*

Company. Though lagging slightly behind, Consett went into steel on what was soon a bigger scale and with the benefits of a fully integrated operation.[14] It proved successful in its new trade.

David Dale announced to the 1884 Annual General Meeting that the demand for steel ship plate was so large as to justify the directors in placing

[14] *CG* 22 Dec. 1882, pp. 977, 986.

The Great Crisis

orders for six more open hearth furnaces. They were still producing more iron than steel plate, but the balance was shifting rapidly. The average price obtained for iron plates, angles, and bars fell by 14s. between January and September 1884. By 1886 over half the Consett output of 72,000 tons of plate was in steel. In the following year their weekly output of steel plate rose to about 1,000 tons. They had eight open hearth furnaces at work in what was to become known as the West Melting Shop. Another eight 25-ton furnaces were then being built in the East Melting Shop. By the end of that year plate capacity had doubled.[15] Others were now turning over to steel, but Consett had gained both a valuable lead and a high reputation for its product. In 1888 the *Vorwarts*, a tanker built for German owners by Armstrong Mitchell at Walker on Tyne, went aground on rocks. Its steel plates, though badly buckled, had, as the German report put it, 'stood up wonderfully'. They had been rolled at Consett.[16] The following year 98.2 per cent of the ships built on the Tyne were of steel. No works was better placed than Consett to supply their needs.

To complete Consett's ability to meet the requirements of shipbuilders, the decision was taken in 1888 to make angles and other sections as well as plate. By 1893 three angle, channel, and girder mills with a weekly capacity of 1,500 tons were at work. To supply them a third, 7-furnace open hearth plant was installed as the North Melting Shop within the angle mill building. Consett weekly steel capacity was now 3,500 tons, or well in excess of the puddled iron capacity of 12 years earlier. The four plate mills' capability was one-third greater than iron plate capacity had been; three-quarters of the iron plate tonnage had already been abandoned.[17] The last puddling furnace went out of production on 1 October 1898 (Louis 1916: 59). Consett Iron was now a major concern, employing a total of almost 6,000 workers, and paying £150,000 a year to the North-Eastern Railway. It had struggled hard to get through the difficult years, but, though the early 1870s still seemed halcyon days, it had remained profitable throughout (see Tables 7.4, 7.5, and 7.6).

---

[15] *Engin.* 12 Aug. 1881, p. 174; *MJ* 23 June 1883, p. 728; *CG* 6 Oct. 1882, p. 545; *I* 19 Sept. 1884, p. 276, 22 Aug. 1884; Newcastle 1887: 14; *CG* 2 Sept. 1887, p. 299.
[16] *ICTR* 2 Mar. 1888, p. 305.
[17] *ICTR* 27 July 1894, p. 119; Consett 1893.

8

The Creation of a Town and a Society—and of the Inertia of Social Overhead Capital

THE construction of the Stanhope and Tyne Railway, the opening of coal and iron ore workings in the coal measures, and the establishment and expansion of iron smelting brought rapid growth to a previously sparsely populated area. Between 1811 and 1841 the population of Conside cum Knitsley increased from 139 to 195, that is, by 40 per cent. In the next 10 years the increase was 1,324 per cent (see Table 8.1). Typically the early Victorians saw such a growth as an improvement on nature. 'Scientific mineral development and a high state of civilisation' had transformed areas that until then had been 'barren wastes'[1] (see Plate 8.1). This transformation was to build an economy and society largely dependent on the continuing operations of the Consett Iron Company. In a more socially conscious age this was to constitute a substantial obstacle to any process of reconstruction which might entail abandonment of the metallurgical industry of the area.

At the 1841 census, 134 of the 192 persons living in Conside cum Knitsley whose birthplace was indicated in the enumerator's returns came from County Durham. In the next five years the Derwent Iron Company erected its 14 blast furnaces, puddling forges, and rolling mills on the moorland around the existing diminutive settlements. In a wider area than the township, the company built 1,300 houses, and speculators ran up many more. The explosive growth of population which accompanied this was largely dependent on an influx of workers from other parts of the British Isles. For many years, as in all rapidly expanding mining and heavy industrial areas, males considerably outnumbered females. For every 100 males there were, in England and Wales, 104.15 females in 1851 and 105.25 in 1861; in Conside cum Knitsley the proportions were 74.76 and 75.95. When this distortion was corrected there was a tremendous potential for population growth by natural increase rather than by further immigration.

The company dominated the economy and the society of both the emerging town and the district. By the end of the forties, in the area as a whole the

[1] *MJ* 15 Sept. 1849.

TABLE 8.1. *Growth of population and housing in the township of Conside cum Knitsley, 1841–1861*

	1841	1851	1861
Population	195	2,777	4,953
Houses:			
Inhabited	34	480	823
Uninhabited	5	57	18
Under construction	—	—	62

population dependent on Derwent Iron Company operations was already reckoned to be 15,000. By the time of the failure of 1857 the company employed almost 5,000 workers. Many were hired on yearly contracts, which were said to have the advantage of ironing out some of the short-term wage fluctuations common elsewhere. As the work-force grew beyond the capacity of the local area to support it by transfers from other occupations, recruiting went on further afield. This was particularly important for special skills: Welsh and Black Country men were especially significant throughout the whole of the northern iron trades. Staffordshire Row was a place-name in the map of Consett indicative of this pattern of migration in the third quarter of the nineteenth century.

In the style of the times the technical and commercial press put the most favourable interpretation on the social accompaniments of industrial development, and on the company's approach to its responsibilities. The working people were said to live in two separate village communities, 'at a sufficient distance from the furnaces to be little affected by the smoke'. The houses, many of which had an upper storey, were in short rows and possessed all 'that relates to decency and comfort'. There was 'garden ground' of quarter of an acre which was offered to all who would cultivate it. Derwent Iron Company established eight day schools, which were supported by its own continuing aid and by 1*d*. a week stopped from wages. Medical provision, involving appointment of four 'well-qualified' men, was financed in the same fashion. The company provided a library and reading-room, 'with fire and candles in winter'. In short, from its early days Consett was growing into a substantial, partly planned, company town. In addition to enlightened paternalism this involved a rather authoritarian regime. The company employed five policemen, whose duties were not only to maintain the peace, but also to report on the state of the houses. If any were found dirty or overcrowded, 14 days notice was given to their inhabitants. The local magistrates acted in conjunction with the company to see that the number of public houses did not exceed what was regarded as reasonable. As a sympathetic commentator observed, the company 'have not exposed the vast capital embarked to the caprice of the crowd, but they began by being thoroughly masters in their own works, and have maintained this mastery by the most legitimate means of scrupulously just,

PLATE 8.1. Consett works and its neighbourhood, 1858

enlightened and able management'. Necessarily, in such a remote district, the company had to continue to play a considerable part in the local communities. As a result, in the 1870s, amid all the discussions about mineral supply, or those centred on whether and when to go into steel, William Jenkins would regularly report to the directors on appeals for schools, or churches, or on individual cases of illness in the community. He continued to live at Consett Hall, was a local JP, and later became an Alderman of Durham County Council. Gradually however the company role declined. There were two main reasons for this, the widening range of provision which was expected, and the growing concern to reduce paternalism by extending local democracy. Some at least thought that there were losses as well as gains from this process. By the 1880s the *Colliery Guardian* was expressing regret that 'localities and work people are insufficiently sensible of the gratitude they owe to the men who are prepared to invest their savings in attempts at mineral development'.[2]

Better communications were needed, for the townspeople as well as for the works. In early 1860 a meeting was held in Shotley Bridge in order to receive a report from the provisional committee on the Newcastle and Derwent Valley Railway. In opening the meeting, the chairman, the High Sheriff of Northumberland, referred to what had already been achieved and to the wants of the area. 'They were in a district which many of the present inhabitants might remember as an uncultured country, with a collection of bad farms, badly farmed. But now the whole country was covered by houses, churches, schools and large manufactories of different kinds. Everything had been done that industry, energy and money could do, and the only thing really wanted was communication with other towns which were in the same position as themselves.'[3] The Derwent Valley Railway was opened in 1867. It was fallacious to suggest, as the High Sheriff had done, that railways were all that was needed; the area lacked a range of the requirements for civilized living. It was gradually provided with more of them.

In the late 1840s, pointing out that they were about 14 miles distant from Newcastle, from Durham, and from Hexham, representatives of the villages of Consett, Blackhill, and Leadgate began a movement for the establishment of a weekly market.[4] By 1860 there was a local newspaper, the *Consett Guardian*. Two years later the Consett Ecclesiastical Parish was organized and there were soon Wesleyan, Primitive Methodist, and Baptist chapels. A police station was put up in 1877. By the mid-nineties the town had a theatre, there was an infirmary supported by the company, and there were monthly county courts, and meetings of the petty sessions on alternate Mondays. A gradual movement occurred towards more democratic local government. By 1865 there was a Local Board and this board of 12 persons remained in control until the 1890s,

[2] *CG* 23 Jan. 1885, p. 140.
[3] *Newcastle Courant*, 13 Jan. 1860, p. 8.
[4] *MJ* 9 July 1859, p. 492.

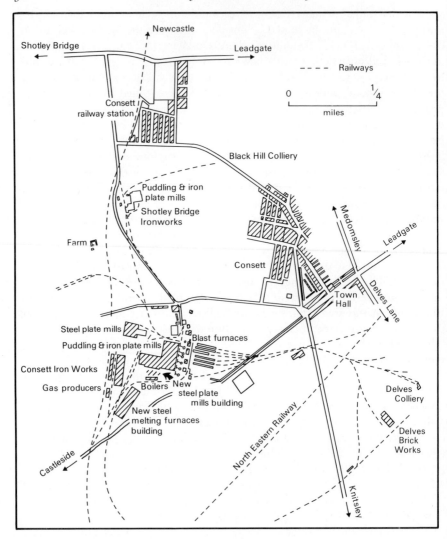

FIG. 8.1. Consett ironworks and town, 1887

when an urban sanitary district was established. Even though important progress was being made in the provision of services and the evolution of local authority control, the economic base, like the urban landscape, remained dominated by Consett Iron (see Fig. 8.1).

Conside cum Knitsley was a completely rural parish before the Derwent Iron enterprise was established. A directory of that time listed its commercial activities as nine farms, one combined farming and milling concern, and one victualler and blacksmith (Parson and White 1828). By the late 1840s, in addition to coal and iron there was some manufacture of firebricks and trade in lead, lime, and timber. Shotley Bridge had flour and saw mills, and at Shotley Grove there were paperworks.[5] The 1860 value of the real property of Consett township was £52,200. Of this, £1,000 was in mines and £34,700 in ironworks. This was the dependency which almost caused the collapse of the mushroom infant community in the troubles from 1857 to 1864. The population at the 1871 census was more or less equal to the size of the work-force employed by Consett Iron at that time. The only other significant employers of labour were R. Dickinson and Co. at Carr House collieries; the South Derwent collieries; the lead mines and smelt mills at Healeyfield to the south-west; a firebrick and a sanitary pipeworks; a bone manure works; and the business of the millwrights and engineers, Robinsons (Slaters 1877). By 1893 the Consett Iron Company employed 6,000 and paid £420,000 annually in wages. The population of the town at the 1891 census was 8,175. At that time 2,700 cottages in Consett, Leadgate, and Blackhill were owned by the company. Much of this property was of poor quality, as for instance in the 'Company Rows', a great block of streets with no amenities in the centre of Consett.[6] The continuing growth of the town—to a population of 9,694 by 1901 and then on to 12,149 twenty years later—remained dependent upon the iron and steelworks.

Outside Consett, too, the activities of the company were building new foci of population. At Langley Park there was no settlement till the mid-1870s; by the mid-nineties a village of 5,000 was grouped alongside the mine, ovens, and brickworks (Mountford 1974: 38, 39). Chopwell contained 4,354 inhabitants in 1901 and 10,000 within a decade.[7] Though some of this growth was dependent on production of coal and coke for sale, most of it was geared to the needs of iron and steelmaking at Consett itself.

The development of such narrowly based local economies created a potentially very unstable employment situation. Already in the nineteenth century there were periods of over-capacity in depressed markets, with high unemployment and acute distress in the small, isolated communities. Both the private interests of the shareholders and the wider interests of society meant that the gathering of a substantial population in a remote location, very largely dependent on one major enterprise, was a constraint on that enterprise's freedom to attempt reconstruction. However, as the 1890s were to show, this did not prevent possibilities of reconstruction being explored. Even more drastic developments were conceivable. In the early sixties a correspondent of the *Mining Journal* had anticipated the consequences of the closure of the ironworks: 'The stoppage of the works would inflict such an evil on the

---

[5] Lewis 1848; *English Cyclopaedia* 1854.
[6] *ICTR* 27 July 1894, p. 120; *Consett Magazine*, June 1958, p. 1.
[7] *CG* 13 Aug. 1920, p. 465, 13 May 1921, p. 1403; *ICTR* 11 July 1924, p. 91, 24 Oct. 1924, p. 675.

district, and also on the creditors, that such a thing is not to be thought of for a moment.'[8] This fear helped rally support at that critical time. Then and more generally the interests of shareholders were well taken care of, but for the wider community the grim forebodings of that correspondent went echoing down the generations, through into the last quarter of the twentieth century.

[8] *MJ* 13 June 1862, p. 412.

9

Consideration of Relocation in the 1890s

In 1859 Tyne Dock was opened on the side of Jarrow Slake. When Consett began to use imported ore most of it was handled there, though some was brought into the Wear. Replacement of the old, low-level Tyne Bridge between Newcastle and Gateshead in 1876 transformed the access to riverside sites suitable for industry above this crossing. The implications were most dramatically illustrated by the commencement of warship construction at the Elswick works of Armstrong Mitchell, but the conditions which made possible the launch of big vessels there also allowed seagoing ships to travel with cargoes to and from wharves on the upper part of the tidal section of the Tyne. It was in this context that Consett Iron began to contemplate developments on the Durham shore of the river. At first these plans involved the more efficient handling of coal and iron ore; eventually consideration was also given to ironmaking, and later still to almost fully integrated operations.

Foreign iron ore could be mined cheaply, and by the late nineteenth century fairly efficient bulk transport by sea had been achieved. On the 150,000 tons of Bilbao ore for which Consett Iron contracted in 1894, the freight charges added only 5s. to the cost in Spain of 12s. a ton.[1] Why not explore the advantage of eliminating further charges on this ore for landhaul by moving ironmaking to tidewater, even if that tidewater was not coastal but merely riverine? Attention first focused on the possibilities of Dunston; later it shifted 1.5 miles up river to Derwent Haugh. Eventually it was to be concluded that any gain from relocation would be marginal. This may have conditioned future Consett responses to consideration of location.

In spring 1889 William Jenkins and David Dale were preparing for a discussion on freight rates with Henry Tennant, General Manager of the North-Eastern Railway. On 25 April Jenkins wrote to his Chairman about arrangements for delivery of ore from the points of importation. He noted that in the early days foreign ore had travelled 23 miles on the line of the old Stanhope and Tyne Railway. After about 1872, the railway company had altered the route for deliveries from Tyne Dock to a 32-mile haul via Washington, Durham, and Lanchester. Charges for railway carriage alone were initially 3s. 2d. This was later cut to 2s. 6d., then to 2s. 2d. and, from 1879, to 2s. 1d. a ton. Additional to these haulage rates were Tyne Dock

[1] CDM 2 Oct. 1894.

wharfage charges of 4*d*. a ton, cranage at about 2*d*., and various other charges which meant that the total transfer cost from the ship hold to the furnace bunkers at Consett was 3*s*. 7*d*. a ton. He reckoned this was equal to about 7*s*. 2*d*. a ton of pig iron or 10*s*. 9*d*. on finished steel plates. It was a 'serious' burden, and therefore saving on it was desirable. The cost of delivery of ship plates to the Tyne and to other leading yards in the North-East was put at 4*s*. a ton. (On deliveries made to the Clyde the charge was 10*s*.) Jenkins admitted that 'much of the above is of course applicable to many other plate makers', but he wondered if they could economize by using a wharf at or near Dunston for importing ore and shipping coal. At least he thought they should mention this when they met Tennant. He was also thinking further ahead: 'There is also the possibility of Consett taking advantage of the erection of steel plant on the river's edge at Dunston and thus diminishing the transport distances of raw and finished material and in this way reducing very considerably the revenue now produced under the present system to the Railway company.'[2] Other coalowners had also contemplated coal exports via Dunston, and in 1893 the North-Eastern itself opened the Dunston coal staithes (Tomlinson 1914: 691, 702). The puchase of the Priestman colliery interests at about this time gave Consett wharfage sites at Derwent Haugh. Their thinking now turned in that direction (see Fig. 9.1).

Derwent Haugh was a hamlet in Winlaton parish with a goods station on the Newcastle and Carlisle railway. By the early 1890s there were coke ovens and firebrick works in the area. For at least the three years 1896–9 Consett management considered the practicality and the economic advantages of a variety of development strategies there. These ranged from using Derwent Haugh for handling raw materials and products, to the erection of blast furnaces, and even to the location of fully integrated iron and steel operations.

In summer 1896 figures were produced which showed that in handling imported ore and in the dispatch of coal, coke, angles, and plate the annual saving through use of Derwent Haugh could amount to well over £26,000 (see Table 9.1). Attention then shifted to consideration of the cost gains from manufacture, starting with iron production for delivery to the melting-shop at Consett. In May 1896 George Ainsworth, long experienced in the blast furnace department, submitted costs assessments in so far as they were above or below those at Consett (see Table 9.2). He commented on these figures:

Most of the items are, necessarily, estimates, but arrived at after careful consideration, and in the case of rates I have taken the same figures that Mr. Holliday assumed in his report.

In order to simplify the statement as much as possible, I give only those items in which there would be a difference between the two series, that is to say where the cost when made at Derwenthaugh would be more or less than when made at Consett. There are in addition sundry minor differences other than those mentioned, but they are so

---

[2] David Dale papers (NRRC 1027/3/31–2).

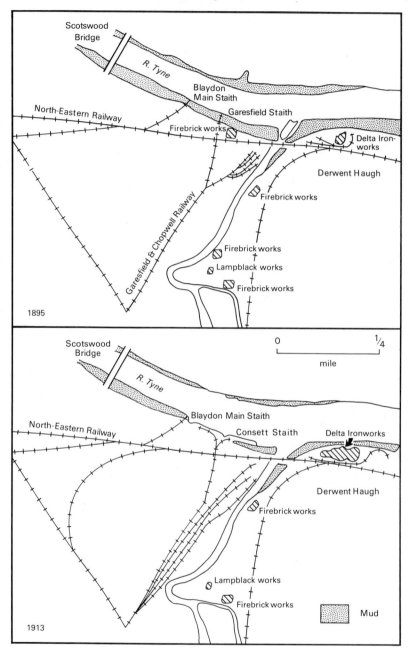

FIG. 9.1. The Derwent Haugh area, 1895 and 1913

TABLE 9.1. *Estimated savings in weekly shipping costs by using Derwent Haugh, 1896*

	Present method			Use of Derwent Haugh			Weekly saving		
	£	s.	d.	£	s.	d.	£	s.	d.
Iron ore (8,000 tons weekly): Via Tyne Dock	1,033	6	8	800	0	0	233	6	8
Coal and coke shipments from Chopwell and Garesfield	n.a.			n.a.			250	0	0
Deliveries of steel plate and angles: To Tyne Yards (720 tons weekly), via Redheugh or Gateshead	141	0	0	117	0	0	24	0	0
Exported via Tyne (310 tons weekly), via Redheugh or Gateshead	60	14	2	50	7	6	10	6	8
Exported via Tees or Hartlepools (600 tons weekly), say half sent via Tyne	72	0	0	56	5	0	15	15	0
Total savings							533	8	4

Source: David Dale papers, 8 June 1896.

TABLE 9.2. *Cost differences for Bessemer pig made at Derwent Haugh for delivery to Consett compared with pig made at Consett*

	Savings at Derwent Haugh (per ton)		Extra costs at Derwent Haugh (per ton)	
	s.	d.	s.	d.
Carriage of ore (Tyne Dock to Consett), 36 cwt. @ 1s. 10½d.	3	4.5		
General blast furnace wages		6.0		
Total savings	3	10.5		
Cost of coke, 20 cwt. @ 6d.				6.0
Carriage of limestone, 9 cwt. @ 1s. 6d.				8.1
Carriage of cinder, Consett to Derwent Haugh				1.8
Carriage of pig, Derwent Haugh to Consett, 20 cwt. @ 1s. 8d.			1	8.0
Slag disposal				2.0
Total extra costs			3	1.9
Net saving		8.6		

TABLE 9.3. *Cost savings in making both iron and steel plate at Derwent Haugh*

Source of savings	Amount (£)
Making 125,000 tons pig iron at Derwent Haugh compared with Consett	17,447
Carriage of this iron to Consett at 1s. 8d. a ton	10,416
Transfer of 94,000 tons of plate to Tyne at 2s. 6d. a ton	11,750
Total saving on 94,000 tons of plate[a]	39,613

[a] Saving per ton of plate would be 8s. 5.14d.

Consideration of Relocation

small that at present they need not enter into our consideration. The only important item as to which I have no information, is the water supply.

The result, you will see, shews that by transferring the Blast Furnaces to Derwenthaugh, there would be an apparent saving of nearly ninepence per ton of pig, but of this amount I am assuming that sixpence would be due to the saving to be looked for in general wages from a completely new plant laid out according to the most economical and labour-saving designs. Obviously this could almost equally as well be obtained at Consett were we prepared to spend the money, and therefore in this sense ought not to enter into the calculation.

On the other hand I have taken the cost of coke at sixpence per ton more at Derwenthaugh that at Consett. Over the last four half-years the cost of Garesfield Coke has averaged nearly eightpence per ton more than Consett, but I have assumed that Chopwell will modify this to some extent. It seems to me, however, that sixpence may not be a sufficient debit as we should be using a Coke which for shipment is fairly well situated geographically, and liberating the same quantity at Consett which is not so well situated either for shipment or the Middlesbrough market.

Ainsworth concluded: 'it is difficult to shew that any saving would accrue by making pigs at Derwenthaugh and sending them to Consett to be converted.' The difference between the cost of making pig at Consett and at Derwent Haugh was reported to the Board as infinitesimal. Ainsworth went on to remark, 'It is clear that if anything is to be done in the way of transference it will require to be a much larger and more comprehensive scheme.'[3] A few days later, in a letter to Dale, he referred to another problem: the site between the North-Eastern main line and the river was rather small for blast furnaces.[4] This was to make any more comprehensive scheme even less viable.

Early in 1899 estimates were prepared of the cost of making pig iron, steel, and plate at Derwent Haugh rather than at Consett. Compared with the costs of iron production at the existing Consett furnaces, the savings on the blast furnace account were still reckoned small. At first consideration, cost reduction in further processing seemed large. It was estimated that 125,000 tons of pig were needed to produce 94,000 tons of plates. The total annual saving on making both pig and plate at Derwent Haugh would be £39,613, or 8s. 5.14d. a ton of plate, on which the average profits of the previous five years had been 18s. 2.50d. a ton; in short, profits on plate would be increased by 46.3 per cent (see Table 9.3).

There were however snags similar in some ways to those which Ainsworth had spelled out three years before. If a new ironmaking plant at Derwent Haugh was compared with a *new* not the *old* furnace plant at Consett, the cost saving per ton of pig iron was cut from 2s. 9.44d. to 0s. 11.98d., a change which would reduce the saving on 94,000 tons of plate to £28,416 a year, or an increased profit of 33.2 per cent. The high capital cost of a new steelworks at Derwent Haugh does not seem to have been allowed for. Clearly it would still

---

[3] G. Ainsworth to Dale, CR 28 May 1896.
[4] G. Ainsworth to Dale, CR 1 June 1896; CDM 9 June 1896.

further pare down the gain in production costs. Additionally, there promised to be yet more site problems. A road diversion would be needed if blast furnaces were to be built. Only 30 acres were available to the company by April 1899. This would be sufficient for a wharf and ironworks, but for steel furnaces and mills a further 44.25 acres would be needed on the other side of the railway. Two landowners were unwilling to give their ready co-operation in making it available, treating the matter, as it was reported to the Consett board, 'in the most acrimonious spirit'.

Very soon after this, attention switched from speculation about Derwent Haugh to the reality of further investment at Consett. The main reason for this was that a new arrangement was made with the North-Eastern Railway. In response to an agreement by Consett to ship all their imported ore through either Tyne Dock or Sunderland, the railway company agreed to cut the delivery charge to Consett from 1s. 11d. to 1s. 6d. a ton from July 1900. This alone was equivalent to a reduction in ironmaking costs at Consett of 9.11d. a ton, or 76 per cent of the calculated saving in operating a new furnace plant at Derwent Haugh rather than at Consett.[5]

This naturally led on to consideration of rebuilding the iron plant at Consett. By 1902 Ainsworth had sent David Dale plans for the replacement of seven blast furnaces capable of 5,000 tons weekly with four furnaces with a total of 8,400 tons capacity.[6] In the event the reconstruction was much less thoroughgoing. Even so, in spurring the company to analyse its costs, and in helping to induce a monopolist railway system to make new freight rate concessions, the consideration of plant construction at Dunston and at Derwent Haugh had yielded valuable returns. Much later Derwent Haugh again featured in Consett development plans, though in a humbler role. In the meantime, now that the question of relocation had been resolved, Consett Iron settled down to continuing production at the original site, where time was to prove that the works had more than half their lifetime still ahead of them.

It is of some interest to put the Derwent Haugh exercise into a wider context both in space and time. Assuming that the 0.4 million tons of iron ore imported annually would make approximately 0.23 million tons of iron, what would have been the cost of assembling this ore and the necessary matching deliveries of coke and limestone at Consett, Derwent Haugh, and at tidewater—say at Tyne Dock? Table 9.4 attempts such calculations. It must be stressed that the freight rates used are no more than estimates, though chosen as objectively as possible.

The Consett Iron Company's calculations in 1896 seemed to suggest that production of 230,000 tons of pig iron at Derwent Haugh might be £53,000 cheaper than at Consett; the estimates in Table 9.4 point to a saving of £45,000. At Tyne Dock the cost advantage would be £23,500. This works out

[5] David Dale papers, 15 Mar. 1899, 29 Apr. 1899 (NRRC 1027/3/31–2).
[6] David Dale papers, 7 Feb. 1902, 26 July 1902 (NRRC 1027/3/31–2).

TABLE 9.4. *Estimated land freight costs of assembling materials to make 230,000 tons pig iron, 1896 (£)*

Material	Quantity required ('000 tons)	Consett		Derwent Haugh		Tyne Dock	
		Charge per ton	Overall cost	Charge per ton	Overall cost	Charge per ton	Overall cost
Ore	416	0.129	53,664	nil	nil	nil	nil
Coke	230	0.008	18,400	0.084	19,320	0.200	46,000
Limestone	104	0.100	10,400	0.175	18,200	0.220	22,880
Total			82,464		37,520		68,880

at £0.102 per ton of pig iron on the assembly cost account, as compared with overall assembly costs at Consett of £0.358 a ton. However it is important to note that, large though this 28.5 per cent freight cost saving was, it represented only a relatively small proportion of the overall cost of ironmaking, and would almost certainly be cancelled out by the costs of building a new plant on a virgin site. In short, around 1900, and as far as Consett was concerned, though certainly attractive on the haulage cost account, the total operating costs of a wholly new, tidewater ironworks compared ill with those of a continuing investment in gradual improvement of the existing works. However, decisions taken for a good reason at one time have ramifying implications for future choices. Thirty years later the enhanced efficiency of ore docks and a very considerable increase in the burdens of land haulage due to a greater increase in railway freight rates than in selling prices for steel meant that the situation had moved more definitely against the inland works. Later again, as the coke rate fell to new low levels, and as giant, specialized ore docks were built, the balance of advantage moved still further in favour of tidewater. When three-quarters of a century after the consideration of development at Derwent Haugh, the main source of coking coal for Consett furnaces shifted from north-west to north-east Durham, the benefits of a coastal location, logistically at least, became dramatically obvious. By that time the longer term implications of choices which had seemed quite rational in the late 1890s could be seen as critical, even as disastrous.

10
Consett, 1890 to 1914

BRITAIN lost its lead in steel output to the United States in 1890, and by 1895 had been passed by Germany. The years from 1880 to the Great War have commonly been regarded by economic historians as ones of British entrepreneurial failure. This has been the viewpoint of a series of authoritative studies from Burn (1940) and Burnham and Hoskins (1943) to Habakkuk(1962), Aldcroft (1964), and Landes (1969). Other (alternative or complementary) factors in the slower growth of British production have been alleged to be the burden of operating more older plant than competitors, and the inhibiting effects of deceleration in the growth of demand. These views have not gone unchallenged, notably by McCloskey. After systematic analysis his conclusion was that 'entrepreneurs in iron and steel, from whatever perspective they are viewed, performed well.' 'Late nineteenth century entrepreneurs in iron and steel did not fail. By any cogent measure of performance, in fact, they did very well indeed.' (McCloskey 1973: vii and 127.)

Whereas some markets for British steel were growing only slowly, often because they were overseas and were vulnerable to competition, sometimes from protected home producers and sometimes from rival foreign sources, the outlets for shipbuilding steels expanded rapidly, were mainly at home, and were only marginally exposed to challenge from foreign sources of supply. It is noteworthy in this respect that McCloskey shows that British open hearth technology (largely concentrated on the production of steel for shipbuilding materials) was more productive than that of the United States up to about 1900, though after that the Americans moved ahead (see Table 10.1).

In both open hearth steel and in plate and angles, Consett Iron, though not the pioneer, was one of the leading and most successful companies of this period. Here if anywhere was a case which confounds the thesis of entrepreneurial failure. It is true that one mark of entrepreneurship may be expected to be efficient location, or response to changing locational forces, and that it was at this time that Consett Iron considered and rejected such a change. However, the relatively small savings this would have brought, the disruption of production and the wider economic and social disbenefits of a relocation to Derwent Haugh together explain why this policy was not followed. In making adverse judgement on the basis of that decision one would be failing to recognize that it is ahistorical to read back into the 1890s the locational lessons which were only unequivocally clear 70 or 80 years later.

TABLE 10.1. *Gross tonnage of merchant vessels built of steel added to the United Kingdom register, 1880–1913*

Year	Gross tonnage ('000 tons)	Year	Gross tonnage ('000 tons)	Year	Gross tonnage ('000 tons)	Year	Gross tonnage ('000 tons)
1880	38	1889	874	1898	1,005	1907	1,191
1881	71	1890	913	1899	1,165	1908	610
1882	128	1891	909	1900	1,111	1909	763
1883	156	1892	922	1901	1,137	1910	1,143[a]
1884	122	1893	735	1902	1,135	1911	1,804
1885	185	1894	835	1903	958	1912	1,737
1886	191	1895	778	1904	1,031	1913	1,932
1887	352	1896	788	1905	1,211		
1888	614	1897	687	1906	1,431		

[a] From 1910 the gross tonnage represents total tonnage launched. Effectively this was all steel tonnage, for in 1909 only 7 thousand tons of the tonnage built were not of steel.

Source: Lloyds Registry.

TABLE 10.2. *Consett Iron Company capacity 1894, 1902/3, and 1911/12*

	1894	1902/3	1911/12
Iron & Steel Works (weekly tons):			
Ironworks	5,250	5,250	7,800
Steelworks	5,300	6,000	7,000
Iron Plate	500	—	—
Steel Plate	2,500	2,830	3,100
Angles	1,500	3,600	3,950
Guide Mill		350	350
Coal & Coke (annual '000 tons):			
Coal	1,000	1,500	over 2,250
Coke	500	600	n.a.

Sources: ICTR 27 July 1894, p. 119; *Trans. Institute of Mining Engineers*, 24 (1902/3), 592, ibid. 41 (1911/12), 420.

There remains one further general comment. McCloskey traced a decline in the relative standing of British open hearth technology after 1900. Although quite exceptional peaks in British merchant shipbuilding were reached in the last three years before World War I, as far as steel construction was concerned the growth in the years after 1900 is necessarily less spectacular than that in the previous 15 years, the period in which steel was replacing iron as the main material used in construction. It is also interesting to see that there is at least some evidence that Consett Iron's growth and success alike both slackened after the turn of the century. The company continued as a noteworthy operation, but its momentum was falling away.

Throughout the quarter-century to 1914, Consett remained one of Britain's leading suppliers of shipbuilding materials. Output continued to increase,

TABLE 10.3. *Dividends paid by Consett Iron Company and other North-East Coast iron and steel concerns, 1900–1912 (%)*

	Consett Iron	South Durham ‚Steel and Iron	Cargo Fleet	Dorman Long
1900–1		10		8½
1901–2		10		6
1902–3	25	nil		4
1903–4	25	nil		nil
1904–5	25	nil		nil
1905–6	27½	12½	5	5
1906–7	40	10	nil	7½
1907–8	33⅓	5	nil	6½
1908–9	20	5	nil	4
1909–10	22½	5	nil	5
1910–11	33⅓	10	nil	6
1911–12	45	20	nil	7½

Source: *Stock Exchange Yearbook*, 1913.

though after the angle mills were brought into production the growth was relatively small (see Table 10.2). Technical efficiency remained high, and there were such wide margins between costs and selling prices that this, combined with the success of its mining and coke-making operations, enabled the company to pay exceptional dividends (see Table 10.3). Yet the turn of the century has been characterized by the company itself in a brief history as 'uneventful'.[1] To the extent that this was true, it was unfortunate, for this was a period of dramatic change in the metallurgical world as a whole, though by no means so much in Britain. Those who did not advance were relatively speaking falling back. There is ample evidence that in the national context Consett remained highly competitive; there is also at least an undercurrent of suggestion that it was to some extent living on its past reputation for high achievement.

Having for the previous 25 years or so 'produced perhaps a larger quantity of excellent shipbuilding material than any establishment in the country', Consett began the 1890s in spectacular style. In 1891 Bolckow Vaughan, with the benefits of a superb location on the Tees estuary, iron ore mines in the hills just behind, Spanish ore mines, and Durham collieries and coke ovens, declared a dividend of 6 per cent. For that year Consett Iron paid 31 per cent. In the summer Bolckows decided that current trading conditions did not warrant an interim dividend; during the same month Consett announced a profit for the year of £276,000. Almost three years later they were singled out as 'a very bright exception to the universal depression and misfortune of iron making enterprises in the north'.[2]

---

[1] *Consett Magazine*, Aug. 1957, p. 14.
[2] *I* 1 Jan. 1892; *The Times*, 3 Aug. 1891, p. 8, 26 Aug. 1891, p. 7; *ICTR* 12 Jan. 1894, p. 46.

At this high point in the company fortunes William Jenkins retired. In 1893 he wrote a survey of the Consett operations which was in some ways a summary of the results of his stewardship (Consett 1893). By the end of that year he was unwell and had largely withdrawn from active management, leaving that to George Ainsworth, a man he had chosen and trained. On 2 August 1894 Jenkins, now in his seventieth year, wrote to David Dale to offer his resignation. He thanked the directors for their 'kind consideration' during the past year of his indisposition, and 'for all the great courtesy shown to me during the quarter of a century I have laboured in your interests as General Manager of the Consett Ironworks'. He died in the following spring. The directors then placed on record 'their appreciation of Mr Jenkins' great ability, sterling character and devotion to the Company's interests ...'.[3]

During the 1890s competition became keener for British steel firms in both domestic and world markets. Over the 10 years from 1889, the last year in which Britain was the world's leading steel producer, output in Britain increased by 36 per cent, in the United States by 214 per cent, and in Germany by 195 per cent. At home new competitors were making their way into trades in which Consett had been so successful. In 1890, for instance, the Dowlais company brought its new steelworks and plate mill into production at Cardiff, and the Frodingham Iron Company commissioned an open hearth plant which marked its entry into steelmaking. Frodingham had early success in finding outlets for its angles and structurals in Midland markets previously dominated by north-eastern producers.

As competition stiffened, some old-established northern firms failed. This happened to two important interior operations. The Darlington Iron Company had been the only wrought iron producer in the region bigger than Consett. Like Consett, Darlington Iron reconstructed as a steelmaker, though without its own pig iron production. It was wound up in 1896. Three years later, when it was liquidated, the value of its plant, land, scrap, and the subsidy which it had received from its old competitors together amounted to less than £30,000.[4] The troubles of the Weardale Iron and Coal Company works at Tudhoe were even more indicative of the way in which competitive conditions had changed. It was equipped with much new plant in the mid-nineties— open hearth furnaces, a cogging mill completed in 1893, and a plate mill which, when set to work in 1896, was claimed as the largest and most powerful in England. The following year Weardale went further, describing Tudhoe as 'the most advanced steel manufacturing and plate mill of its kind in Great Britain if not in Europe'—certainly a gross exaggeration. Within four years it was decided that so keen had the competition become that it was necessary to move the whole of the steel and plate operations from Tudhoe to a new site at

---

[3] *ICTR* 12 Jan. 1894, p. 46; David Dale papers (NRRC 1027/3/31–2) and CDM 18 June 1895.
[4] *ICTR* 31 July 1896, p. 150; *CG* 21 Apr. 1899, p. 714.

TABLE 10.4. *Consett finished steel capacity and the quantity of materials needed to support it, 1905* (tons per week)

	Finished products	Steel ingots	Pig iron	Coke
Plates	2,900	4,257	3,405	3,745
Angles	1,900	2,267	1,813	1,994

Source: Report by G. Ainsworth, 5 Sept. 1905.

tidewater on the estuary of the Tees at Cargo Fleet. Announcing this, the Weardale chairman, Sir Christopher Furness, reckoned that the only alternative would have been to remodel the works on the most modern lines.[5] The threat to Consett was that, despite its efficiency, in the new, still more strenuous struggles for business, location might eventually tell against it in the same way as with Darlington and Weardale, not to mention the once-great works in the hills of South Wales. Although it continued to be profitable, in some respects Consett's management now seemed irresolute, even diffident, in contrast to the way they had given a lead to the region only 20 years before.

By this time Consett was in some respects an ill-balanced works (see Table 10.4). As the General Manager, George Ainsworth, pointed out, their weekly plate capacity could be increased by 100 tons given a larger supply of steel and iron.[6] Already by this time some British companies were engaged in what was then commonly called the 'Americanization' of their operations. With a view to the most effective reconstruction of Consett operations, Ainsworth and his Chief Engineer, James Scott, made a tour of American works. No comprehensive redevelopment programme resulted from this, but attention focused on the desirability of improvements in their ironmaking. In the 1900 report the directors recommended that £100,000 be added to the blast furnace reconstruction fund, and in the following year they commissioned the consultant F. C. Roberts of Philadelphia to report on their needs. The practical outcome was slight. Only one new furnace was built, and that not before 1908. Even then, because of bad trade conditions, it was not blown in for another two years.[7] Meanwhile, since the commissioning of the angle mills in the early nineties, finishing capacity had not increased greatly. Apart from the new blast furnace the main thrust of development in this period was in coal and in coke. Here too there were some signs of irresolution.

Most of the Consett Iron production of coal had come from pits near to the town. However, in the years around the turn of the century the expansion was either in the Browney valley or in the high ground north and west of Rowlands Gill. In the Browney Valley, Langley Park colliery was extended in the 1890s.

---

[5] *ICTR* 9 Oct. 1896, p. 488, 14 May 1897, p. 718; *CG* 19 Jan. 1900, p. 123, 1 Nov. 1901, p. 967.
[6] G. Ainsworth report, CDM 5 Sept. 1905.
[7] *CG* 3 Sept. 1900, p. 228; *ICTR* 2 Jan. 1903, p. 28.

TABLE 10.5. *Coal production by certain outlying collieries, 1895 and 1899* ('000 tons)

	Langley Park	Garesfield	Chopwell	Total
1895	198	161	—	359
1899	347	157	98	602

Source: David Dale papers (NRRC 1027/3/31–2).

By 1904 it operated five shafts; a decade later over 400,000 tons of coal were being raised a year. Developments northwards towards the Tyne had been contemplated for many years. As early as 1881 the company had requested the North-Eastern Railway to make a line to join the Newcastle and Carlisle at Riding Mill, with a point on the Blaydon and Consett Line between Ebchester and the station at Lintz Green. It pointed out that such a route 'would pass over a most valuable unworked coalfield'.[8] The railway was not built and further development in the area had to wait for a decade. Consett had long owned royalties north of the Derwent, but large-scale development began only after the purchase of the Garesfield colliery and of associated railways and Tyne shipping facilities from the Priestman family. The Derwent Haugh to Garesfield railway was extended to High Spen in 1891 and then on to what was known as the Chopwell New Winning. In the latter part of the nineties Chopwell colliery was brought into production, though it as yet remained a small factor in Consett Iron's coal production (see Table 10.5).

Considerable conservatism and uncertainty was shown by Consett Iron in relation to coke manufacture and particularly to the question of by-product recovery ovens. To some extent this may be excused, for they had an abundant reserve of coal suitable for production of first-rate metallurgical coke in the traditional beehive ovens, but they were definitely far from pioneers of carbonization practice at this time. Pease and Partners had built the first British by-product ovens at Crook in 1882. By the mid-1890s the most enterprising of the Teesside ironworks had begun to install them next to their furnaces. In 1896 Consett received reports on the by-product ovens operated by Pease and by Bernhard Samuelson and the Carlton Iron Company. A year or so later, when they built ovens at Chopwell, they opted for beehives. Up until 1902 they continued to discuss and, as it seems in retrospect, to vacillate over the question of by-product ovens.[9] After this they did decide to install them at the Templeton cokery; their capacity there was doubled in 1912. Langley Park was provided with by-product ovens in 1915. It was not until a few years later that the company showed a pioneering spirit in coke technology.

[8] Mountford 1974: 15; CDM 28 July 1881.
[9] CDM 1 Aug. 1896, 4 Nov. 1902.

By World War I, Consett Iron was a highly successful producer of coal, coke, and steel products. It was then reported as raising annually 2.25 to 2.5 million tons of coal (apparently an indication of productive capacity rather than actual output), and its 612 beehive and 180 by-product ovens made respectively 376,000 and 178,000 tons of coke. There were seven blast furnaces dating from the 1870s and one newer, bigger furnace. At full work their annual iron capacity was about 320,000 tons. Steel ingot capacity was 330,000 tons. Four mills could roll up to 150,000 tons of plate, and the three section mills were capable of 100,000 tons (Louis 1916: 60). Thus there had been no further increase in finished steel capacity since Ainsworth had reported about the material balances of the works in 1905. In iron and steel, Consett, unlike so many of its competitors, was a single plant operation. It was wholly dependent on imported ore, and had its own ore interests overseas. The large physical outputs were still matched by high profitability.

11
Unravelling the Puzzle of Consett Success

THE discovery of the Cleveland Main Seam was explicitly recognized as presenting a serious threat to the interior works of the North-East. One by one both the existing works and also plants later established in the interior failed in a sad sequence of outright collapse, slower rundown, or removal of operations to more favourable locations—Ridsdale, Hareshaw, Tow Law, Witton Park, Ferry Hill, many of the works of the Darlington district, and the major venture at Tudhoe. Yet forty years after the ore was proved at Eston, when Bolckow Vaughan, the main beneficiary of that discovery, paid a dividend of 2.5 per cent, Consett declared one of 33.33 per cent. As early as 1883, when its shares were at a pronounced premium, it was described as 'the most remunerative of our iron companies' (see Table 11.1). Three years later, at a time of deep depression, Consett Iron £7. 10s. shares were standing at £18.[1] It was almost universally highly regarded. In the mid-90s a leading trade journal remarked: 'To the up-to-date steel manufacturer and engineer, not to speak of the shipbuilder and general customer, the history, location and capabilities of Consett are as well known as the road to Mecca is known to the devout followers of Islam.'[2] A little later the *Joint Stock Companies Journal* wrote, 'Consett has a prosperous record with few to compare with it in England, Europe or even the United States.'[3] The material evidence for this is that in the 50 years to World War I its average annual dividend was 23 per cent. Why did Consett Iron succeed so well?

A careful review of the factors involved was published by H. W. Richardson and J. M. Bass in 1965. The analysis which follows owes much to their account, though it fills it out in some respects and takes issue with some of the conclusions. Surprise about Consett's achievement centres on the apparent disadvantages of its location and site. This leads on to examination of the way in which supply of raw materials and distribution of finished products was organized. Process cost savings may enable a firm to capitalize on any advantages it possesses in location, raw materials, or marketing. Alternatively—and, given man's inclination to inaction when things seem to be going well,

---

[1] *The Times*, 17 May 1886, p. 11; *Engin.* 26 Oct. 1883, p. 390.
[2] *ICTR* 31 Jan. 1896, p. 149.
[3] Quoted by Richardson and Bass 1965: 71.

TABLE 11.1. *North-East steel shares, 1883*

	Nominal value (£)	Ruling price (£)
Consett Iron	10	23 to 23.5
Bolckow Vaughan	20	20.125 to 20.25
Darlington Steel and Iron	10	1.0 to 1.50
Palmers Shipbuilding and Iron	35	27.25 to 27.75

Source: ICTR 26 Oct. 1883, p. 511.

probably more likely—such savings will permit it to counteract to some degree any corresponding disadvantages. Technology and scale of production are relevant considerations here. The relationship of the firm with other producers is of importance—is there co-operation to divide the market? Finally, and the activating principle of it all: how good is the management?

Nineteenth-century contemporaries were well aware of the unsuitability of Consett's location. As one of them observed, it 'appears very odd' that Consett Iron should be paying much higher dividends than Bolckow Vaughan, as 'a slight examination of the two would suggest that Consett was much less favourably placed'. Another recognized how strong the pull of tidewater had become: 'It is too soon to be seen in what part of the North East . . . growth of the trade will be known, but generally it must be accepted as a fact that the tendency will be towards the seaside.'[4]

Yet the problem of location deserves fuller analysis, for it is clear that by assiduous attention to it Consett managed to reduce very materially the burdens which it implied. In the early years extensive use of higher-grade ores from Cumbria both reduced the need to haul poor Cleveland ore inland and at the same time enabled the company to produce better-quality products. These in turn gave increased margins over the costs of the earlier stages of manufacture and eased Consett into markets more secure than those for rails. From the 1870s the main movement of raw materials was of Spanish haematite railed from the lower Tyne. Richardson and Bass suggest that in the mid-1880s the charge for this haul was 1*d.* per ton mile. In fact there is evidence that the rate for the whole journey was gradually reduced from 3*s.* 2*d.* to 2*s.* 1*d.* a ton; even so, with handling charges included, the overall rate by the 1890s was 3*s.* 7*d.* a ton, or about 1.34*d.* per ton mile. It was, Jenkins said, a 'serious' burden. Surprisingly, the rate on finished steel railed to the port seems to have been lower. In the eighties it averaged 0.8*d.* a ton mile (Richardson and Bass 1965: 73, 74). Consett enterprise and the willingness of the North-Eastern Railway to accommodate its main customer in north-east Durham ensured that these land hauls and substantial charges were not allowed to bring Consett Iron down.

---

[4] *MJ* Feb. 1889, p. 364; *CG* 24 Apr. 1881, p. 613.

In its consideration of the other main raw material, coking coal, the Richardson and Bass analysis seems to be deficient. It is implied that other firms in the region were as well placed as Consett in relation to good supplies. For reasons of geology and of location this was not so. The north-west Durham plateau, within which Consett lies, is geologically speaking at the node of the Northern coalfield. Here is the greatest concentration of seams of the highest coking quality. It is true that there was also good coking coal in south-west Durham, and indeed the Crook area became the greatest focus of coke manufacture in Britain, but this area was eventually seriously affected by watering. Moreover, it lies some 25 to 30 miles by rail from Teesside. In serving ironworks there from ovens in the Bishop Auckland/Crook area, the North-Eastern was by no means as much in the service of one coal or iron company as it was in that of Consett in the north-west. Consequently, it may be assumed that it was less inclined to give any one company rates on coal and coke as competitive as those which it provided for Consett on iron ore.

Some have suggested that possession of large reserves of coal and revenue from coal and coke sales was more important than iron and steel to Consett's large profits and dividends. MacLaren alleged this in 1905, and repeated it twenty years later.[5] His claim was at odds with the early analysis of Edward Williams, and with later evidence as well. It is true that Consett was able to obtain coal cheaply compared with most of its rivals. As early as the Derwent Iron Company days it was said to be able to mine coal at costs no higher than other concerns were paying for royalties alone (Richardson and Bass 1965: 81). Moreover, Consett coal was of the finest coking quality and was, especially in the early years, almost on the spot for iron smelting. Between 1864 and 1914 Consett coal production tripled to 1.5 million tons; coke output remained at about half that of coal. Half the coke was sold—about 50,000 tons of it went as return cargo in boats which had brought Bilbao ore to the Tyne. The sale of coal and coke provided a useful counterweight to the iron and steel trade, for the periodicities of their two trade cycles were not identical (Richardson and Bass 1965: 81, 82). There is also evidence that in the twenty years before the Great War Consett's fuel sales were growing more quickly than its steel trade (see Table 11.2). Yet the evidence points to the conclusion that it was efficiency in both coal and steel industries which made Consett so profitable. In 1880 for instance, when the company declared a 20 per cent dividend, it was remarked that its success was due in the main to coke and iron, the coal trade in that year yielding only slight profits.[6]

In 1911 there appeared a study of the profitability, over a 13-year period, of the 92 companies which two years earlier had mined 80 million tons of coal, over one third of the national total. Their average annual dividend on ordinary capital, from 1898 to 1910, had been 9.6 per cent. Consett, with an average

---

[5] *TES* 3 May 1905 p. 77 and Aberconway 1927.
[6] *The Times*, 9 Sept. 1880, p. 4.

TABLE 11.2. *Estimated trends in the structure of productive capacity at Consett Iron, 1894, 1902/3 and 1911/12* (index nos.)

	1894	1902/3	1911/12
Pig iron	100	100	148
Crude steel	100	113	132
Steel plate	100	113	124
Steel angles[a]	100	263	287
Coal for sale[b]	100	193	227
Coke for sale[c]	100	150	264

[a] The high index for angles in 1902/3 and 1911/12 reflects the lateness of Consett's entry to this field.

[b] This is an estimate based on a coal to coke ratio of 1:0.65 and the fact that half the coal mined was converted into coke.

[c] Takes account of furnace requirement of 1.099 tons of coke per ton of pig and the fact that roughly half the coke made was sold.

Source: based on Consett Iron Company records.

TABLE 11.3. *Average dividends of leading steel and coal firms, 1898–1910*

Firm	Average dividend (%)
Consett Iron	31
Bell Bros.	28
Guest, Keen, & Nettlefold	12
Weardale Steel, Coal, and Coke	7
Bolckow Vaughan	6
Ebbw Vale Steel, Iron, and Coal	5

Source: Richardson and Walbank 1911.

over three times as high, topped the list. The possible conclusion that, in direct opposition to the suggestions quoted above, it was in fact its interests in steel which raised it above the run of coal companies is put into perspective when it is compared with other concerns straddling both coal and steel (see Table 11.3).

In low processing costs, associated with scale and technology, Consett came to occupy a prominent place. It was one of the biggest steel producers by the end of the century, and moreover, unlike some other concerns, had all this capacity in one works and wholly in open hearth furnaces rather than, as in so many instances, divided between the open hearth and Bessemer processes, the latter of which was becoming increasingly outmoded for most finished products. In plate its only equal at that time was Colvilles, though shortly afterwards the South Durham group also became a major producer. Consett could not be regarded as a pioneer of new technology, but when it entered a trade it did so purposefully and efficiently. Not only did Jenkins examine other

TABLE 11.4. *Shipbuilding in the North-East and other districts, 1889–1920* ('000 tons)

	1889	1895	1906	1920
Tyne and Blyth[a]	306	174	394	371
Wear	217	138	335	319
Tees and Hartlepool	194	211	292	273
Scotland	396	391	642	778
Rest of UK	196	152	234	399
TOTAL	1,309	1,066	1,897	2,140
Tyne, Blyth, and Wear as % of total	39.9	29.3	38.4	32.2

[a] 1889 figure includes Whitby.

works before his firm embarked on steelmaking, but he continued to keep in touch with technical progress elsewhere. There is considerable evidence of the efficiency of Consett as a steelmaker, further refuting the suggestion that profits were dependent on coal. The two great centres for the production of shipbuilding materials were the North-East and mid-Scotland, each serving great regional markets (see Table 11.4). Generally the North-Eastern firms were lower-cost producers.

In the early 1890s they were able to deliver plates to the Clyde at a price said to be 7s. 6d. below the cost of production in Scotland. A decade later, though the charge for carriage to Glasgow was 8s. to 10s., the North-Eastern firms were selling angles there a few shillings a ton below the prices local producers could meet (Richardson and Bass 1965: 75, 76). Although in about 1890 two representatives from Dowlais were surprised to find no new machinery in the plate mill at Consett—they praised the melting shop—Consett was a highly competitive member of the North-Eastern group of producers (Thackerey and Lockley c.1890). This was highlighted by evidence given to the Royal Commission on Labour in 1892: it was shown that, though some of the cost advantage of the North-East seemed to be due to low wages, these only seemed low because they were tonnage rates and reflected the fact that better machinery and organization resulted in lower manning levels than north of the border. Consett at that time rolled and treated plates for 2s. 3d. a ton, whereas in Scotland the same class of work was said to cost 4s. 6d. to 4s. 9d. John Cronin, a representative of the Associated Society of Millmen in Scotland, indicated to the Commission that in a particular works the mill had six to eight men around the rolls for an output of 80 tons of plate per shift. At the same time Edward Trow, who had graduated from puddler to the post of executive secretary of the Board of Conciliation for the Manufactured Iron Trade of the North of England, pointed out that at Consett, with five men at the rolls, they averaged over 100 tons per shift. Also the Scottish works, lacking their own blast furnaces, had to operate cold metal practice. In the early 1890s, while these works were sometimes recording losses, Consett was still distributing handsome dividends. By 1894 it was opening markets for its new angle mills on

TABLE 11.5. *Production and sales of plate by the members of the North-East Steel Platemakers Association, 1905* (tons)

	Production	Deliveries in North-East
Consett	147,448	102,630
Bolckow Vaughan	41,675	32,823
Palmers	2,340	1,404
South Durham	280,236	199,890
John Spencer	17,560	17,650
TOTAL	489,259	354,307

Source: Minutes of meetings of North-East Steel Platemakers Association.

the Clyde and trying to establish trade with the two main Belfast yards of Harland and Wolff and Workman Clark, even though they lay in the natural market area of the Scottish mills. The liberal margins which efficiency gave Consett Iron were well indicated in the year to June 1900. Consett's average production costs in that period for plates and angles were £4. 12s. 11d. and £4. 4s. 6d. respectively. Selling prices were £7. 14s. 0d. and £7. 11s. 6d.[7]

As Richardson and Bass indicated, Consett never held a monopoly, even within the region. From 1897 it was a member of the North-East Steel Platemakers Association, which gave all the producers a share of the work available. The evidence suggests that Consett regarded the association as a convenience which it could employ to bolster a position which would still have been extremely strong without it. In 1905 Consett made 30 per cent of the plate output of the North-East, and about 11 per cent of the national total of some 1.3 million tons. Though it sold almost 70 per cent of its plates in the North-East area as defined by the associated producers, its share of their total deliveries in this district was a little below 29 per cent[8] (see Table 11.5).

In terms of wage costs Consett Iron was in a favourable situation. It dominated the local labour market, and therefore, when necessary, could squeeze down the rates of pay with little fear that it would lose its men. This is not to imply that the company was harder in its attitudes to labour than rival concerns, but is merely a reflection of one of the benefits of a location which in other respects was disadvantageous. Into the mid-1870s Consett workers seem to have been paid higher wages than their counterparts in Middlesbrough. At that difficult time, when Consett was sometimes the only works in the region on full time, Jenkins pressed for wage cuts. There were other forms of pressure. For example, during the winter of 1879/80 a new incline was put up to the blast furnaces. The improved handling which it made possible enabled the company to serve dismissal notices to 30 men. Yet notwithstanding such a heavy hand on its workers in the name of efficiency, in the same operational

---

[7] Royal Commission on Labour, 1892, pp. 373–9, evidence of Cronin and Trow; CR.
[8] Minutes of meetings of the North-East Steel Platemakers Association.

year the company paid a 20 per cent dividend, and in addition resolved to distribute a bonus from accumulated profits and from the reserve fund.[9] Jenkins was said to be fair with the men, but it is clear that on the labour front, the Consett Iron Company was most certainly not running a liberal commonwealth in north-west Durham.

It must be stressed that for many years Consett competitiveness benefited from low overheads. In 1864 the new Consett Iron Company paid just over £295,000 for plant having a capacity of 150,000 tons of pig iron and 40,000 to 50,000 tons of finished iron. Shortly afterwards it bought the Shotley Bridge works for £55,000. In 1872 it invested only £50,000 for its quarter-share of the Orconera Iron Company, which after the first few shaky years was not only a convenient, but also a profitable associate. In short it was not plagued by the over-capitalization which dragged down some iron concerns. Necessarily, with the passage of time and with new calls on capital, the advantages associated with this cheaply acquired plant were reduced. There was large outlay for the new plate mills and steelworks in the early eighties and for the angle mills and extended steel capacity a decade later, though they were all built in times of uncertainty or of depressed trade when construction costs would be lowered. Thirty years later still, it was to be a deliberate choice to press on with another and major reconstruction of plant during deep depression. Long before this it had been recognized that Consett's success was due to much more than low capitalization. In 1877, referring to the initial low purchase price, an authoritative source commented: 'It is mainly owing to such an excellent start that the company have all along been able to pay such good dividends, although not a little credit is due to the excellence of the management.'[10] The latter factor was increasingly recognized as of prime importance.

Its coal reserves, the scale, technology, and general efficiency of its iron and steel plant, and the initial cheapness of the whole enterprise were important assets of the Consett Iron Company, but the whole had to be co-ordinated and activated. All in the end depended on the quality of management. In the 55 years from 1864 the company had only three General Managers—Jonathan Priestman for the first five years, William Jenkins from 1869 to 1894, and George Ainsworth, who was originally chosen by Jenkins as blast furnace manager, but was then appointed to succeed him, and who was to remain in control to beyond the Great War. Priestman seems not to have been a great success, as is suggested both by the independent witness of Edward Williams's two 1869 reports on operations and by Jenkins's own opinion of his predecessor. Under Ainsworth profits were on average at their highest, but it was in the quarter-century of William Jenkins's control that the firm foundations were laid of a Consett tradition of organizational excellence. It is

---

[9] *The Times*, 8 Jan. 1876, p. 6, 4 Dec. 1877, p. 4, 6 Mar. 1879, p. 7, 11 Feb. 1880, p. 9, 9 Sept. 1880, p. 4.

[10] *JISI* 1877, pp. xi, xii; Richardson and Bass 1965: 82–4.

perhaps more accurate to describe the management of Consett in this key period as the achievement not of one man but of a partnership—Jenkins, the ironmaster and expert in the iron business, and David Dale, the financial adviser and company chairman. It was a powerful combination of talents. Dale had wide commercial experience and contacts, in railways, in shipbuilding, and in the arbitration service of the region's iron trade. His involvement with Consett extended over almost 50 years. He certainly had no doubt of the exceptional qualities of their General Manager. As he wrote of his success in persuading Jenkins to come to Consett: 'never had a director of a company done a better day's work than he did that day' (Richardson and Bass 1965: 79). Jenkins proved to be an excellent co-ordinator of the factors of production— liberal yet firm with his men, a good judge of talent in others, and keen to keep up-to-date technically so long as profits could be increased by so doing (Richardson and Bass 1965: *passim*).

It seems likely that Consett's management, to a far greater extent than that of any of its rivals, benefited from singleness of purpose. Certainly the conditions for the exercise of that quality were less propitious elsewhere within the region. Palmers were operating in two trades, shipbuilding and steel, and were unable to devote their undivided attention to either. Bolckow Vaughan for many years conducted business at three works, and even when they closed the plants at Witton Park and Middlesbrough and concentrated their attention on Eston, they suffered from the fact that since they became a limited concern in the mid-1870s they had been controlled by outside capitalists, largely from the Manchester area. Bell Brothers was a regional concern, but dissipated its efforts. Port Clarence was by far their biggest operation, but for a number of years they also made iron at Walker. They tried the manufacture of aluminium, had chemical interests at Washington, and from the early 1880s were distracted by involvement in salt and then in alkali production on Teesside. The Furness group, which took shape at the turn of the century, undoubtedly had purpose and drive, but also a number of distinct operations, of varying levels of efficiency, at the Hartlepools, at Stockton, at Tudhoe, and within a few years, also at Cargo Fleet. Even Dorman Long, a new, vigorous concern, suffered an inevitable increase in complexity and dissipation of effort as it grew, by acquisition and extension of existing works, from the original nucleus of the Marsh and Britannia mills through subsequent association with Bells and later with operations at Redcar and the interests of the North-Eastern Steel Company and Samuelsons. By contrast with these concerns, with all of which it was to a greater or lesser degree in competition—Consett Iron's attention was focused on a single integrated plant, though direct sales of coal and coke were also important. In the winter of 1869 William Jenkins had set out the aims of his regime: 'to keep capital entire' with the hope of 'remaining on the spot to which they were attached'. The company's achievements owed much to the fact that both he and his successors maintained a steady course towards that goal.

That the active principle in Consett commercial achievements lay in the sense of purpose and the quality of its management was acknowledged by contemporaries. Their success had been won after 'one of the most chequered histories among English manufacturing establishments' and against the tendency towards the 'seaside'. Self-sufficiency in coal and extra income from the sale of surpluses in that section of the business, and the benefits of low standing charges were all recognized, but special reference was made to 'very good management' and the conduct of the business in 'a most energetic manner'. As a consequence, the works 'by rare judgement and capacity' had been 'brought to a point of commercial efficiency rarely attained'.[11] No stronger proof could be found that business success is above all the reward for managerial excellence, rather than the result of especially favourable external factors. Most interesting of all—though one of the unanswerable 'ifs' of history—is the question whether, in the absence of Consett's locational disadvantages, management of such abilities would have been brought into being.

[11] *MJ* 21 Aug. 1880, p. 962, 6 Aug. 1887, p. 975; *ICTR* 31 Jan. 1896, p. 149.

12

The Great War

IN operations and in profits, Consett Iron had an undistinguished war, but the period was in a number of respects a major turning-point in the company's fortunes. By this time its leading position in shipbuilding steels, established in the 1880s and consolidated during the following decade, was slipping. The substantial wartime expansion of national steel capacity, particularly in the shipbuilding field, benefited its rivals. Even more important, a new, more fully co-ordinated form of development planning for the whole of the British industry now began to take shape. It took many years to mature, but as it did so, there emerged a livelier concern for optimization of locations for production. Such an emphasis could only act to Consett Iron's long-term disadvantage.

Financial results for 1914/15 showed a profit of only £247,000, the worst for 20 years. Of the succeeding war years, only 1915/16 produced higher profits than in 1913, and they were far behind the 1900/1 figure. The company was short of labour at critical times: workmen enlisted in such large numbers that in 1916, when shipyards were crying out for material, one of Consett's largest plate mills was idle for a time. Prices of raw materials, stores, and timber for the pits, as well as wages, all increased very sharply. At the same time maximum prices were fixed. The coal-mines were brought under the administration of the Coal Controller, the steelworks within the aegis of the Ministry of Munitions. Much more significant than these inconveniences, which affected all producers to a greater or lesser extent, was the fact that there were no large-scale wartime plant extensions at Consett, but major ones elsewhere.[1]

Consett Iron completed extensions to the staithes at Derwent Haugh in 1914, and in the following year commissioned the new battery of by-product coke ovens at Langley Park. There were no important projects on the steel side. Meanwhile the needs of war brought government requests to the industry for expansion, first mainly in the provision of shell steel, and then above all in the production of shipbuilding material. This government persuasion helped reinforce the development planning which a number of companies were already contemplating. For instance, early in 1914 the Steel Company of

---

[1] 'A Short History of the Consett Iron Company', *Consett Magazine*, Aug. 1957.

Scotland had circulated its shareholders with a recommendation that they should take a half-share with the Frodingham Iron Company in the works of the Appleby Iron Company, which adjoined the latter, with the intention of building an open hearth steel plant and plate mills there. The outbreak of war frustrated this scheme, but towards its end the project was taken up by the new United Steel Companies of which Frodingham became a major part. In the twenties this development was to be carried through to completion. During and just after the War, with government encouragement and financial assistance, new plate mills were installed in Scotland and in South Wales. There were also important extensions to capacity within the North-East.

In autumn 1912 the South Durham Steel and Iron Company had planned a new plate mill for the Malleable Works, Stockton, which would be designed to roll wider, thinner plates. During the War it came forward with a much more ambitious project, a completely new, fully integrated Teesside works. This scheme did not receive the necessary government approval, but it was a clear indication of the continuing expansionist thinking of the South Durham group. Later the Furness interests, which dominated both South Durham and Cargo Fleet, secured an associated outlet for plate and angles in the new shipyard at Haverton Hill. Other big Teesside developments also directly threatened Consett's market share. By November 1916, the Bolckow Vaughan board was committed to the expenditure of an estimated £225,000 on a new plate mill at Cleveland Works. Earlier that year, Dorman Long embarked on the conversion of a newly acquired, small blast furnace plant at Redcar into a major integrated works. By the end of the year they had agreed with the Government to install a plate mill there.

By these and other large developments the pressures to be anticipated in the postwar shipbuilding steels markets were increased. The members of the North-East Coast Steelmakers Association had delivered 468,000 tons of plates and angles to home shipbuilders and marine engineers in 1913. In the same year, the Scottish Steelmakers Association delivered 395,000 tons. Five years later, the Scottish association reckoned that when extensions then underway were completed, their capacity would be 56 per cent above the 1913 level.[2] To accommodate this increase and the extensions then being made in the North-East as well as in Lincolnshire and South Wales would require comparable extensions in the shipbuilding outlets. Major expansion of shipbuilding capacity was indeed underway, but it remained to be seen whether the building ways of new and extended yards could be filled on a regular basis. On that would depend the profitability of the steel extensions and the competitive situation in a major part of that heavy steel sector on which Consett was wholly dependent.

In the latter half of the Great War, Government attention was turned to the

[2] Report of the Departmental Committee on Shipping and Shipbuilding Industries, Cd. 9092 (1918), pp. 27, 41.

prospects for British manufacturing in the post-war world. During 1917, as part of this review, a Departmental Committee of the Board of Trade held an inquiry into the state of the steel industry. Committee members included Hugh Bell, a member of the boards of Bell Brothers and of Dorman Long, and Benjamin Talbot of South Durham. One of the avowed aims was to make the British industry competitive with American steelmaking. Evidence was taken from highly respected operators and from consultants. Generally the proceedings of the committee can have provided the controllers of Consett with little comfort or encouragement.

One of the main witnesses was Charles G. Atha, who had worked in the United States until 1901 and was now General Manager of the Frodingham works. Between March 1916 and January 1917 he had been back to the USA, largely spending his time visiting works in connection with the supply of shell steel. He was immensely impressed by what he saw and returned with convictions that the British industry must concentrate to survive. A minimum size for a commercial steelworks (as opposed to those in special trades such as armaments, engineering, or quality products) should be 10,000 tons a week. Atha reckoned that currently only one British firm met that target; no more than nine of the 59 'commercial' producers were more than half as big. Consett was certainly well below Atha's recommended minimum. He advocated radical action to improve the situation. For tinplate bars, rails, billets, shipbuilding material, and other standard material, 'I cannot see any reason why the trade of this country should not be concentrated in a comparatively few efficient steelworks such as I saw in America'. The change would be dramatic. 'My conclusion, therefore is that, as previously stated, the work to be done to reconstruct our industry on modern lines is enormous, as the final aim must be nothing short of a complete replacement of the great majority of existing plants by very much larger and more efficient units, and learning to manage and operate such plants in accordance with modern practice and methods.' The new works would be on greenfield sites. He favoured use of home raw materials. It was left to another witness, Axel Sahlin, to take up his case for new works, though at the same time Sahlin stressed the need for foreign ore. For neither the Atha nor the Sahlin prescription could Consett be seen to be a suitable growth-point.

Axel Sahlin was an iron and steel consultant whose peacetime headquarters was in Brussels, and he had operational experience in the industry both in Britain and overseas. He condemned the expansion policy being followed under the pressure of war demands: 'Instead of patching old plants, I recommend concentration and combination and new construction.' The true foundation on which to build was the ore supplies, and these should be from overseas. His vision had extremely wide horizons: 'In Brazil lie, awaiting European enterprise, the largest fields of high-grade ore of which I have any knowledge.' For the location of the industry within Britain, Sahlin drew the logical conclusions. 'In a country geographically situated as Great Britain (and surrounded by sea) it would seem to be an axiom that iron and steel plants should be located at a deepwater port, so that raw materials and products could be delivered directly from works yard to ship or vice versa.'[3]

The report of the Departmental Committee was published in 1918. It envisaged an annual steel capacity of 15 million tons. Such a target was extremely ambitious, being 7.3 million tons in excess of the record pre-war output of 1913, and 5.3 million tons above the wartime peak reached in 1917. It was also to prove greater than the output of any year before 1949. With the avowed aim of making this expanded capacity competitive with that in America, the Committee took up the arguments used by Atha and Sahlin. It is true that their idea of plant size was still conservative, but they spelled out the need for a new structure for the industry, new patterns of supply, and, by implication, new locations.

This means the laying down of complete new units, blast furnaces, coke ovens, steel furnaces, rolling mills. Modern plants must have a minimum capacity of 300,000 tons per year, if they are to work economically. In turn these plants must be fed from commensurate resources of coal and iron ore.

These resources should, as far as possible, come from within the Empire:

To develop these resources large sums must be expended, not only in the working of the mines, but also for the construction of railways, the building of special ships, for the transportation of ore, of large docks and discharging and loading appliances. Secondly the plants themselves must be designed on such a scale that continued expansion is easily possible and expansion should take the form of the filling out of a predetermined design, and not of successive accretions to existing plants.[4]

This was the first occasion in the history of the British iron and steel industry on which a general development scheme was outlined. It was therefore the first time the inherent difficulties of Consett's situation were, by implication if not explicitly, contrasted with ideal patterns of production. Consett Iron was just about big enough to be viable under the new thinking, but it had certainly grown piecemeal. More worrying still was the suggestion that the proposed large-scale operations, set within a network of rationalized raw material flows, should be located at specially designed ore terminals. In the past, competitors within the North-East had generally been much less successful than Consett. Most of them had current deficiencies or problems, but for Bolckow Vaughan, Dorman Long, South Durham and their associates at Cargo Fleet, it was clearly possible to overhaul, to demolish and reconstruct, and so to approach much nearer to the ideal held out by the Departmental Committee. Consett could not follow the same path without adding removal to modernization. The challenge to its viability was not spelled out specifically at

---

[3] Minutes of Evidence before the Iron and Steel Industries Committee: C. Atha, 28 Mar. 1917 and 4 May 1917; A. Sahlin, 24 Mar. 1917.

[4] Report of the Departmental Committee appointed by the Board of Trade to consider the position of the Iron and Steel Trades after the War, Cd. 9071 (1918), p. 20.

this time. Even if it had been, the unhappy implications of the report could scarcely have been clearer.

Unfortunately for British steel's overall competitiveness—though luckily for Consett—expansion in and immediately after the war did not follow any logically structured, grand strategy, but instead a policy which Sahlin had characterized and condemned—'I think that the putting up of steel furnaces right and left is making a beginning at the wrong end.'[5] In September 1922, by which time war-induced extensions and the expansive thinking of the immediate aftermath of war had both come to an end, Walter Layton, who had been a member of the Munitions Council, and was now a director of the National Federation of Iron and Steel Manufacturers, considered why developments had fallen so far short of the ideal. At the same time he to some extent endeavoured to explain away the necessity for radical change. As he put it, the extensions

> were planned with a view to rapid construction rather than to efficiency of production, and in order to enlist the full resources of the steel industry and its personnel they were widely scattered. Every firm which had an available site and reasonable facilities for dealing with materials was encouraged to go ahead. If the Ministry had had a single eye to the question of cheapness of production, it is probable they would have preferred a scheme of two or three self-contained super-plants, very carefully chosen as regards geographical position and probably situated at ports. The case for doing this would have been particularly strong in Great Britain, for we have, thanks to our long metallurgical history, an abnormally large proportion of old plant, much of it situated at comparatively unsuitable places. But any such considerations were over-ridden by the imperative necessity of speed. And against the arguments just set forth we should bear in mind that what we call American methods of bulk production, though suited to war conditions when enormous quantities of a single standard article were required, are not so well adapted to the peace-time trade of Great Britain, which is of a much more retail and miscellaneous character than that of the United States. (Layton 1922: 489.)

In adopting such a view, Layton seemed willing to condemn the British industry to occupy a place in a small-time league. Even so, ideas of a more efficient, larger-scale economic order of things, and even the concept of a national steel industry, implying some sort of overall planning, had been given an airing. Development thinking would never be quite the same again. Consett had escaped the difficult prospect of competing with a series of 'super-plants' as Layton had called them, but in the course of a few years the steel capacity of the industry had been increased by roughly 50 per cent, so that unless the post-war boom proved to be a permanent feature, the strength of the competition which any plant would have to face would be stronger than before the war. Very early in the new decade it became clear that the boom years had passed.

---

[5] Minutes of Evidence before the Iron and Steel Industries Committee: A. Sahlin, 24 Mar. 1917, pp. 31–2.

13

The 1920s: A Critical Decade in Consett Development

ALTHOUGH it had not been involved in large new construction projects, Consett Iron came out of the Great War with what seemed to be a large and well-managed integrated works. In finished steel it was concerned with fairly standardized products, which it turned out in what were, by British standards at least, large tonnages. As Tolliday has pointed out it benefited in the boom conditions of 1919 and 1920 from the possession of cheaply mined coal which enabled it to keep its costs down when prices were bounding upwards (Tolliday 1987: 49). In these respects it was favourably placed in an industry whose deficiencies, long recognized, were becoming more and more critical. Some years before two foreign observers had summarized these problems, problems which were becoming painfully obvious to almost all by the 1920s:

Orders individually too small, intermittent manufacture, endless changing of rolls, inability to specialise plant, difficulty in writing off rapidly, a lessened interest in adopting the most modern mechanical improvements, and notably continuous mills and corresponding shipping facilities, continuous heating furnaces, electrical charging machines, etc. (Billy and Melius 1904: 20, 21.)

A number of these features had since then been improved, but the indictment was still widely applicable.

Consett, despite its past successes, current profitability, and high reputation was by no means free of these defects. Its works were in some respects ill-balanced, and contained a good deal of obsolescent plant. In the twenties they had to operate in an increasingly competitive situation, initially due to the effects of wartime extensions at home, but soon also resulting from keen trading by overseas rivals, in Europe often helped by plant rebuilt following war damage or destruction. Conditions were to be worsened by depression in demand, but it is noteworthy that even in the three years of good trade, 1918, 1919, and 1920, Consett profits averaged no more than 2.2 per cent above the 1913 level (see Table 13.1). In the new circumstances, location was again to become an important issue in Consett planning. A final complication was that at this critical time there occurred another change in Consett's top management. George Ainsworth, General Manager for 26 years, died in the middle of the post-war boom.

TABLE 13.1. *Consett Iron Company profits, 1899–1922* (before charging depreciation or for capital spending)

Year	Profit (£)	Year	Profit (£)	Year	Profit (£)
1899	459,770	1907	542,833	1915	247,140
1900	699,506	1908	428,459	1916	657,962
1901	611,127	1909	289,407	1917	552,812
1902	320,997	1910	291,599	1918	540,950
1903	250,286	1911	370,415	1919	512,292
1904	266,175	1912	451,660	1920	899,142
1905	267,953	1913	636,582	1921	506,430[a]
1906	356,848	1914	482,731	1922	102,446

[a] In 1921 there was a 4-month stoppage at the collieries which affected the coke ovens and the works. The average annual profit for the 23 years to June 1921 was £462,742.

Source: *The Times*, 16 May 1922, 21 June 1923.

Coal-mining and manufacture of coke remained of great significance in Consett success. Their profitability provided a large financial support for reconstruction in steel; at the same time they helped to localize that reconstruction. In 1927, repeating an assessment made 20 years earlier, Lord Aberconway wrote of Consett Iron, 'its large dividends are understood to have been due less to its success in steelmaking—for which it is not so well situated as works on the coast—than to its fine coal royalty.'[1] This opinion paid less than fair attention to the quality of the Consett works, and came ill from an ex-chairman of Palmers, but there can be no doubt of the importance of coal and coke. The care which Consett Iron devoted at this time to extending that part of its operations is sufficient witness to this.

By 1921 6,900 men worked in Consett collieries. Three years later, though national output was over 7 per cent less than in 1913, Consett employed in eleven collieries 2,500 more men than eight years before. A few years later still, annual coal capacity was put at 2.5 million tons or 0.5 million more than the average of the three years before the War.[2] Expansion on the fuel supply side culminated in the company's construction of new coke ovens, a venture of a pioneering quality fit to stand with the initiatives taken under William Jenkins.

For years after the Great War British carbonization practice was, generally speaking, antiquated. In 1930 a visiting Belgian delegation reckoned that 60 per cent of the ovens were obsolete, and another critic suggested that only two or three of the 30 coking plants in County Durham were not 'obsolete by American and German standards'.[3] Before the war, though a major producer of coke, Consett had by no means been a pacemaker in the switch from beehive to by-product ovens. Now this situation was dramatically changed. In December 1920 the decision was taken to build a new plant to carbonize 6,000

[1] Aberconway 1927: 184, footnote; *The Times*, 16 May 1922, p. 22.
[2] *ICTR* 11 July 1924, p. 91, 24 Oct. 1924, p. 675; *Colliery Engineering*, Mar. 1933, p. 82.
[3] *Econ.* 25 Jan. 1930, p. 133.

tons of coal weekly. The Fell coke works were in operation by March 1924, (Plate 13.1). After close study of American practice, which, after a very slow start in comparison with continental Europe, had surged ahead with major wartime extensions, Consett Iron was the first British concern to introduce the narrower style of slot oven, and to substitute silica bricks for conventional refractories in its construction. Resolutely they went further, sinking the Victory pit nearby to supply much of the coal which Fell would need, and building a works at Templetown to make the bricks from Butsfield ganister. Their initiative, maintained through difficult times, received wide commendation. In 1924 a continental authority on refractories, Dr Endel of Berlin, described Templetown as the world's best-equipped refractory brickworks, and their product as its finest. Twelve years later the British carbonization expert, R. A. Mott, recalled the wider 'demonstration effect' of the venture: 'The value of the example to the British industry, in which in recent years too few pioneering steps have been taken, was enormous.'[4] Consett had other, older-type by-product coke plants at Templetown and Langley Park, but between 1927 and 1929 erected yet another, completely new cokery in the Derwent Haugh area—though 1.5 miles inland from the riverside site considered by the company for ironmaking 30 years before. In this instance contracts were obtained for sales of gas to the Newcastle and Gateshead Gas Company. By the end of the twenties the four Consett by-product plants could carbonize about 29,000 tons of coal a week. Chopwell, though a relatively recent development, was never provided with by-product coking. Generally the new investments in coke manufacture, involving not only increased efficiency and the recovery of tar, ammonia, etc., but also the recovery of gas which was used in the iron and steel departments, helped to maintain steelmaking in the same location.

Notwithstanding its history of commercial success, and its new achievements in coking practice, Consett works was ill-placed to meet the competitive challenges of the 1920s. Its equipment was old at a time when much new plant was being commissioned; and, when the trend to larger-scale operations was pronounced, it still ran a number of small units (see Table 13.2). Radical reorganization in ironmaking, contemplated twenty years before, had still not been carried through. The newest furnace, commissioned in 1910, was 75 feet high, 12 feet 6 inches in hearth diameter, and capable of about 2,500 tons of iron weekly. The other seven furnaces, dating from between 1873 and 1880, were only 55 feet high by 10 feet in diameter. By the early twenties one of them had been improved to make as much as 1,200 tons a week.[5] There were three melting shops, with 39 open hearth furnaces, whose heat capacities ranged from 20 to 35 tons. For a considerable time other

---

[4] *ICTR* 18 July 1924, p. 113; Mott 1936: 103; SMT 2/73, Report of Belgian delegation of Aug./Sept. 1930.

[5] *ICTR* 22 Apr. 1927, p. 641.

Plate 13.1. Fell coke works in the early 1950s

TABLE 13.2. *Units operated by Consett Iron Company, 1922*

Units	Number	Production ('000,000 tons p.a.)
Collieries	8	up to 1.9/2.0
Coke ovens	n.a.	0.35
At Consett works:		
Blast furnaces	8	
Open hearth steelworks	2	
Plate mills	4	
Section mills	3	

Source: Consett Iron Company Prospectus, *The Times*, 16 May 1922.

companies in the region had been operating bigger units. For instance in 1915, when the average furnace size at Consett had been 29.3 tons, it was 31.1 at Bolckow Vaughan, 52.6 at Dorman Long, and 67.7 tons at South Durham—though it must be remembered that the 'Talbot'-style furnaces installed at South Durham were always of a bigger nominal capacity than non-tilting furnaces (*Rylands* 1915). The plate and section mills were now definitely aged, a substantial impediment in a trade in which there was soon to be a surfeit of newly built capacity. These indicators of lagging technology were recognized by management, and assessments for a general reconstruction, estimated to cost £1.5 million, began immediately after the ending of hostilities. As had now become general practice, consideration was given to the best foreign methods.

Visits were made to American works, and in autumn 1920, when a new blast furnace and melting shop was contemplated, a development report was considered by the Board.[6] Reconstruction was then delayed by difficulties accompanying introduction of the 8-hour day, fears of possible nationalization of coal, and by general industrial uncertainty. On the other hand, pressure for modernization increased as other heavy steel firms commissioned new mills. Dorman Long's first plate mill at Redcar began rolling in 1920; a second unit followed in 1922. The major extensions in Scotland came into production. In 1918 construction was started on the new Appleby works. In this instance completion was frustrated by the 1921 slump, but it was clear that eventually Scunthorpe would be a low-cost producer in plate, as it had long been in sections. Appleby mill was not finished until early 1927, having been helped by a £650,000 guarantee by the Government. Long before that the necessity for Consett to cut costs had become clear. Already by late summer 1922 it was being estimated that British ship plate capacity was half as great again as in 1914, but demand was well below the pre-war level and still falling. Only efficient producers could make money—perhaps only they could survive (see Table 13.3).

[6] CDM 5 Oct. 1920, 2 Nov. 1920, 7 Dec. 1920.

TABLE 13.3. *Mercantile shipbuilding and ship plate prices in Britain, 1913 and 1919–1929*

	Tonnage launched ('000 tons)	Ship plate price[a] (July)		
		£	s.	d.
1913	1,932	6	7	6
1919	1,620	18	5	0
1920	2,055	27	0	0
1921	1,538	15	0	0
1922	1,031	10	0	0
1923	645	10	5	0
1924	1,439	10	5	0
1925	1,084	9	2	6
1926	639	7	17	6
1927	1,225	8	7	6
1928	1,445	8	7	6
1929	1,522	8	12	6

[a] The price given is of plate in Scotland; it provides an overall index.

Sources: for tonnage, Burnham and Hoskins 1943: 286; for prices, Carr and Taplin 1962: *passim*.

Looking back, Consett's Chairman, Clarence Smith, summed up his board's dilemma.

In 1919–1920 we had a very old steel-melting and plate and section rolling mills plant, which had been worked to its full capacity for many years—but during the war years especially—without opportunity for maintenance and repairs. This plant could not possibly operate to compete with other manufacturers in this country apart from any question of competition with continental material. The company was therefore faced with either allowing this plant to go into disuse—when it would have been left with, as its main operating departments, the collieries and coke ovens only—or to take the bold policy of demolishing the old steelworks and rolling mills and putting down in their place a modern plant.[7]

There were however more barriers to progress with reconstruction. The two most important were the continuing distraction of possible involvement in mergers, and the problems posed by uncertainties of trade.

Many amalgamations were proposed after the war, though few of them came to fruition. As early as 1918 Consett Iron was approached concerning involvement in a consolidation. In spring 1921 and again two years later it was invited to consider purchasing the uncompleted Appleby works, or, on the second occasion, 'taking an interest' in both Appleby and in the Frodingham works, by lending its name to that of United Steel Companies in raising capital. In 1921 it was agreed that Clarence Smith and the General Manager should visit Scunthorpe to obtain more information; in 1923 the Board was

[7] *The Times*, 19 June 1926, p. 19.

resolute enough to decide that the company's commitments were such as to rule out the proposal.[8] Late in 1922 they were invited to take an interest in Whitwell's works, which would have brought them, though only in a small way, into iron manufacture on Teesside. The Consett Board resolved that 'in the present circumstances' they should not do so. (In the middle twenties they were left out of some of the wider-ranging merger proposals. They were not included in the discussions which Lord Pirrie had in 1923/4 with a number of major firms throughout the nation, nor in the group of six with which Henry Bond, as President of the National Federation of Iron and Steel Manufacturers, carried on negotiations.[9])

As far as trade conditions were concerned, there were more uncertainties. In the booming business lasting into 1920 there was a natural reluctance to embark on rebuilding when profit margins were high and when, at the same time, prices for new plant and equipment were inflated. When recession began and rapidly deepened into depression the opportunity for redevelopment seemed for a time to have passed. Eventually it was recognized that reconstruction costs had also been dramatically reduced, at a time when the existing plant could not have been profitably at work.[10]

Even in 1919 and 1920, when demand was buoyant, although Consett's finished steel capacity had been considerably greater than in 1914, its output had been less. This anomaly was attributed in part to the deficiencies of railway services.[11] By the Annual General Meeting of August 1921 foreign competition had already become crushing. Home pig iron was priced at £7 a ton, a level said to be below the costs of production; at the same time foreign iron was coming into the Tyne at £4. 10s. Girders made in Britain were priced at £16. 10s., but some imported material was being delivered on the North-East Coast for £10. In mid-October prices for home-produced finished steel were cut by £3. 10s.; three weeks later, to save an estimated £1,000 a week, Consett Iron reduced the wages of steel melters by 45 per cent.[12] Prospects remained bleak in 1922. It was in September that W. T. Layton of NFISM (National Federation of Iron and Steel Manufacturers) made the chilling observation that whereas British shipbuilding was unlikely for some time to consume even as large a tonnage of ship plate as before the war, national capacity to roll such plate had increased by about 50 per cent (Layton 1922: 489). It was in this darkling setting that Consett reconstruction got underway.

By summer 1921 the West Melting Shop had been demolished and not only was its site almost cleared, but preparatory work had begun for the installation

---

[8] CDM 24 Jan. 1918, 5 Mar. 1918, 5 Apr. 1921, 6 Mar. 1923.

[9] CDM 5 Dec. 1922. The Pirrie scheme involved Dorman Long, the Furness group, and Bolckow Vaughan on the North-East Coast; Colvilles in Scotland; and Baldwins in South Wales and the Midlands.

[10] *ICTR* 5 Aug. 1921, p. 183; *CG* 5 Aug. 1921, p. 394; Carr and Taplin 1962: 381.

[11] *CG* 13 Aug. 1920, p. 465, 6 Feb. 1920, p. 397.

[12] *CG* 12 Aug. 1921, p. 464; CDM 1 Nov. 1921.

TABLE 13.4. *Shipbuilding and plate production in 1920 and 1921* (tons)

Year	Shipbuilding tonnage commenced	Plate sales for all purposes
1920	2,396,819	1,611,700
1921	568,438	648,600

Note: at this time an estimated 1 ton of steel was required for every 2.1 gross tons of shipping built.
Source: Economist, 11 Feb. 1922, p. 221.

of new plant. By early 1922 two considerations persuaded the company to speed up the reconstruction. Firstly, completion of new rival mills sharpened the competition. Secondly, general depression had by now swept away most of the market so that there was relatively little to lose by being out of the business for a short time if the result would obviously be beneficial in the longer term (see Table 13.4).

In January 1922 no vessels were launched on the Tyne; the month in which that had last happened had been 39 years earlier. Two months later, speaking to his board, Smith noted that from 1918 to June 1921, notwithstanding the relatively low output levels noted above, they had made record profits. A good deal of these had been retained and were now available for the reconstruction programme. He went on to stress that this should be undertaken as quickly as possible, for 'it appears that we are so self-contained that we ought again, with a reconstructed plant, to have opportunities to make money on iron and steel.' It was scarcely a clarion call, but Smith stressed that the slump in steel demand gave them the opportunity to rebuild 'at the least cost possible in these times, and to get ready for the large demand for iron and steel production that is sure to come eventually'.

They now planned a new melting shop, and also a sheet mill 'to utilise saleable scrap'. By May some figures were available: there would be a very small outlay at the blast furnaces to increase output from 5,000 to 7,500 tons a week, a new melting shop capable of 4,000 tons, and two plate mills costing £863,000. The following month it was decided to go ahead with the steel plant, but, hoping to economize in capital outlay, the Board asked their manager to look into whether it would be possible 'to purchase at a reasonable figure one or more existing plate mills, which were known to be in the market for sale'. Yet at the end of that year, recognizing that trade was likely to be bad for a long period, the directors resolved to press ahead. All units except for the blast furnaces would be closed and would either be replaced with new plant or, as in the case of the angle mills, undergo a thorough reconstruction.[13]

The main period of activity began in July 1923. By ordering in depressed conditions, costs for new equipment were reduced by approximately one-third

[13] CDM 7 Mar. 1922, 2 May 1922, 13 June 1922, 9 Aug. 1923.

from the levels of 1920, and engineering expenses also were cut. Indeed, competition among major suppliers to secure contracts was so keen that Davy of Sheffield was said to have lost £50,000 on the construction of one of the new plate mills. Naturally it was anticipated that the new installations would improve Consett's competitiveness, and the Board were determined not to be constrained in maximizing the benefits. Accordingly, in September 1924 they gave formal notice of an intention to leave the North-East Coast Steelmakers Association, 'so that the Company might have freedom in making up a book for the commencement and working of the reconstructed mills'.[14] Events were to justify these expectations. The melting shop gave them 9 open hearth furnaces of 70 tons heat capacity. The new slabbing mill was at work by August 1925, and the 9-foot 6-inch plate mill in the following month. The development programme was virtually completed by the end of that year. It immediately lowered Consett's costs, giving it an edge over competitors. Before the end of October the Board of the South Durham Steel and Iron Company was told that the new Consett mills had forced down the price for ship plate delivered to North-Eastern yards to £7. 5s. 0d., a figure well below even the depressed general price. South Durham could not match their rival's costs, so that 'this price will undoubtedly leave an actual loss.'[15] Consett now managed to wean away some important South Durham customers including Armstrong Whitworth and Swan Hunter and Wigham Richardson. However, the Gray yards in West Hartlepool rejected its approaches. Two years later it was reported that Consett was still winning orders for plates and sections 'much in excess of its proportion'.[16]

The rebuilding of Consett works made the company an effective competitor, but could not make it a great commercial success in the terrible conditions of the twenties. Its financial decline came later than that of most of its rivals, but for over 10 years from December 1924 it was unable to pay a dividend on ordinary shares (see Table 13.5). In 1927, largely as a result of the wider effects of the 1926 coal strike, Consett Iron lost £78,000. In 1928 and 1929 profits were £300,000 and £320,000 respectively. Altogether between 1919 and 1931 the company spent £5.5 million on plant, £3 million of it coming from reserves. Both before and after this expenditure Consett was a more efficient producer than the main Teesside firms, though not generally than the South Durham/Cargo Fleet group (see Table 13.6). As with Fell coke works, it was warmly commended for its enterprise. Here was a situation radically different from much of the British steel industry as viewed by an American critic in 1926, an industry in which 'older companies with slack directors have been content to see their industries decay and have endeavoured merely to save as much as possible by working the older plants until they become obsolete'. At

[14] CDM 30 Sept. 1924; *Steel Review*, 8 (1957), 44.
[15] South Durham Steel and Iron Company minutes, 28 Oct. 1925 (NRRC 1066/13/1).
[16] Tolliday 1987: 53; CDM 8 July 1927.

TABLE 13.5. *Dividends on ordinary shares paid by Consett and other North-Eastern iron and steel companies in the 1920s*

Year	Consett	Palmers	South Durham	Cargo Fleet	Dorman Long	Pease and Partners
1920	10	12½	—	5	10	—
1921	4	2½	10	5	5	—
1922	3¾	nil	10	nil	nil	—
1923	7½	nil	10	nil	nil	8
1924	2½	nil	10	nil	nil	1½
1925	nil	nil	10	nil	nil	nil
1926	nil	nil	5	nil	nil	nil
1927	nil	nil	5	nil	nil	nil
1928	nil	nil	6	nil	nil	nil
1929	nil	nil	6	nil	nil	nil
1930	nil	—	6	nil	nil	—

Source: *The Stock Exchange Yearbook*, 1932.

TABLE 13.6. *Profit ratios of North-East Coast steel companies, 1919–1929* (trading income divided by year-end book value of net assets)

Year	Consett	Dorman Long	Bolckow Vaughan	South Durham/Cargo Fleet
1920	17.1	11.4	17.5	n.a.
1921	10.2	4.9	9.6	n.a.
1922	−3.0	1.9	−0.8	n.a.
1923	4.9	2.3	−4.0	10.7
1924	8.4	4.2	−0.4	8.0
1925	3.5	1.9	−1.5	5.4
1926	−2.3	−1.4	−0.4	−0.6
1927	−0.6	2.2	2.3	5.1
1928	5.0	2.9	2.7	7.5
1929	5.2	3.6[a]	[a]	7.1

[a] In 1929 Bolckow Vaughan was merged into Dorman Long.
Source: Tolliday 1987: 26.

the beginning of the Consett scheme a leading trade journal had referred to it as 'the most important steel works reconstruction carried out in this country for many years'. After its completion, the same journal was unstinting in its praise: 'theirs is the faith which accomplishes when the faint-hearted would fail.'[17] However, notwithstanding such well-earned endorsements, the programme was by no means an unqualified success when considered in a wider context. It exhausted the company finances when depression of trade and sharpness of competition meant that margins of profit were narrow. As a result, though far-reaching, the reconstruction was not complete, and Consett operations remained ill-balanced. Most important, it preserved a production

[17] *ICTR* 10 Aug. 1923, p. 191, 1 Jan. 1926, p. 8; *Iron Age*, Autumn 1926.

TABLE 13.7. *Major Consett Iron Company developments in the 1920s*

Date	Development
1921–2	Fell coke works and ancillary projects
1922–5	Consett works reconstruction:
	new 42 in. slabbing mill
	new 9 ft. 6 in. plate mill
	new 6 ft. 6 in. plate mill
	reconstructed section mills
1927–9	Derwent Haugh coke ovens

location whose long-term viability was at best highly questionable and at worst untenable.

In spite of its large-scale rebuilding, Consett Iron remained a rather small and unevenly equipped plant (see Table 13.7). In 1927, in a presidential address to the Iron and Steel Institute, F. W. Harbord recognized that American ideas of plants of a minimum weekly capacity of 10,000 tons of heavy products were inapplicable to British conditions, but he argued for specialization on one class of finished products and for what he called 'balance' between various departments. On the subject of balance, while mentioning no plants by name, he observed:

In my opinion a very common, if not the most common defect in iron and steel plants, is the lack of balance. It is not unusual for so much money to be spent, often far more than is necessary, on one part of the plant, that it is impossible, for financial reasons, to carry out the original intention of modernising the plant as a whole. We find plants in which the open hearths and mills are of latest design, equipped with the most modern devices, while the coke ovens and blast furnaces are insufficient in number to meet the requirements of the steel plant, or are old and working under conditions which make it impossible to produce pig-iron economically. (Harbord 1927: 711.)

With the important exception of his reference to coke ovens, Harbord might have been thinking of Consett when he made those comments—though there were many other candidates for his strictures. Having spent £4 to £5 million on rebuilding, Consett still needed to find another £1 million if it was to complete the process by modernizing ironmaking. However, by 1927 Lloyds Bank was pressing the company for care in financial matters. As another critic of the industry acknowledged, Consett could not make the desirable further outlay 'until the £4 to £5 million had been digested'.[18] The onset of the slump in the autumn of 1929 meant that this process of digestion was cut short. Meanwhile, paradoxically, the question of the unsuitability of the Consett location again came to the fore.

[18] Vaizey 1974: 49; Bruce-Gardner, 12 Nov. 1930, in SMT 2–72.

Consett had been one of the pioneers in the introduction of efficient delivery of overseas ore. Now, although there were no major new works, a number of firms were taking up the recommendations of the Board of Trade Departmental Committee for expansion of iron and steel capacity adjacent to ore terminals. In 1921/2 Brasserts, the internationally respected steel consultants, are said to have recommended that Consett should follow this model in its redevelopment, though it must be stressed that the Consett Board minutes give no hint of such a suggestion. During the next few years, the need grew for careful thought about location as a result of the new, larger scale of plant, because of higher costs of operation in established and often congested sites, and because of rises in transport costs. Older, urban locations were suffering from the burden of rising rates brought on by the increased responsibilities of local government. Consett Iron paid £29,000 in local rates in 1914; in 1927, £118,000.[19] Dramatic increases in rail freight charges inevitably penalized locations like that of Consett most of all. Even at South Durham, Lord Furness had shown that from July 1913 to October 1921, whereas the delivered price of steel sections had increased by just one-third, the rise in the costs of carriage of raw materials and finished products had averaged 112.9 per cent.[20] An inland plant would be even more badly affected by such changes, and indeed the adverse impact of freight rate changes on their own operations were reported to the Consett directors as early as January 1920. Yet despite this apparently strong argument for coastal location to minimize costly land hauls, there were also important disadvantages in the development of greenfield sites. Even by that time the cost of rounding out an existing works was much less than that of a completely new one. The capital necessary for a greenfield site works had been increasing rapidly. For Britain generally a fully integrated plant was estimated to cost for every ton of capacity for finished products from £4 to £4. 10s. in 1890, £6. 10s. to £7 by 1912, and as much as £14 in 1934 (though these figures are for slightly different products in the various assessments).[21] Consett Iron made its choice in what seemed to the outsider a quite unequivocal way, but there is at least one indication that the decision to remodel the existing works rather than look elsewhere was a much closer thing. In late 1923 E. J. George gave evidence to the Railway Rates Tribunal on the flat-rate increases in mineral traffic charges, rises of a type which naturally hit hardest the firm with haulage of a large number of minerals, and particularly one with longer hauls. He was reported as saying:

in 1921 his Company had to decide whether they should reconstruct the Consett works. He had considered whether, instead of doing that, he should transfer the works to another site, where he would not have to use the railway except for finished products. The circumstances seemed to be in favour of reconstruction at Consett, but his

[19] *Iron and Steel Industry*, Nov. 1927, p. 49.
[20] *Econ.* 3 Dec. 1921, p. 993.
[21] Harbord 1927: 35, 36; Burn 1940: 509n.

views would probably have been altered if he had thought that the flat-rate was to be a fixture.[22]

At the end of the twenties, after a decade of depression, reconstruction and amalgamation were widely mooted. Consett was again approached for possible involvement. At least two proposals concerned 'downstream' development: in 1928 Sir John Noble suggested that Consett Iron might take over the old Armstrong Whitworth engineering subsidiary of A. and J. Main; eight months later they were invited to assume management of the forge and spring shops at Newburn on Tyne.[23] Neither suggestion was taken up. More important were the attractions of association with other steel firms. From early 1927 the Consett Board occasionally discussed the evolving merger situation on Teesside—at least once the proposed merger considered taking in Lincolnshire steel as well. The Teesside developments included the successful linking of Dorman Long and Bolckow Vaughan, and then the protracted and ultimately unsuccessful discussions for the inclusion of the South Durham/Cargo Fleet group. In midsummer 1929, before the merger with Bolckows, Arthur Dorman had asked the Consett General Manager whether his company would be interested in participating. The Consett minutes recorded that 'Mr George had explained that Consett was practically a self-contained concern, and whilst there did not appear to be any advantage gained for Consett from such an arrangement, if a proposition were put to them they would no doubt give it consideration.'[24] The new links went ahead without them; the reconstruction and rationalization which followed were to build up more competitive complexes in favourable locations. Meanwhile, in spite of depression, of their own uncertainties and second thoughts, the rebuilding programme in the twenties had given Consett a new lease of life as an independent concern. Retrospectively this may be reckoned a not unmixed blessing. It laid the foundation for further commercial success, but in the process also perpetuated an increasingly eccentric location. To have changed this in the twenties would not have been easy; later it was to prove impossible.

[22] *ICTR* 28 Dec. 1923, p. 968.
[23] CDM 1 May 1928, 8 Jan. 1929.
[24] CDM 6 Aug. 1929.

14

The Thirties: Depression, and National and Regional Organization

IN the 1920s, a decade of difficulties in British basic industries, opportunism had made Consett Iron one of the pioneers of modernization in steel. Nationally 1929 was a year of record steel output, but the British industry as a whole could not yet claim to have overhauled itself. In September an editorial in a leading trade journal stressed that much hard work lay ahead: 'times have changed, and in the face of developments in reorganisation and concentration of production in European countries, and to a lesser extent in the United States, our present practice is untenable, and only by adopting similar methods can we hope to survive as an important iron and steel producing nation.' By contrast, the same editorial recorded impressions from a recent visit to Consett. There the works 'leave little to be desired'; the company's expenditure seemed to have been amply justified.[1] During the early thirties the steel firms were pressing for tariff protection and at the same time groping towards some form of effective national organization which would include provision for co-operative development planning. It did not prove an easy process. A 1933 PEP report recognized that 'The abolition of opinionated decisions, narrow horizons and bias is as important as the scrapping of old plant and abandonment of inefficient techniques. Fortunately, the industry is one which has more reputation for common sense than stubbornness.'[2] That was a good summary of the problems, but, as events were to show, was far too sanguine about rational reactions by those in positions of leadership. In the middle and later years of this decade investment in physical reconstruction spread much more widely through the industry. A consequence of this was that some of the margin of lead which the pioneers of rationalization had gained was eroded. Inevitably, in a more generally modernized industry, the inherent disadvantages of Consett would again be apparent.

Soon after the collapse on Wall Street in October 1929 demand for steel in Britain began spiralling downwards. At the same time foreign producers, faced with similar difficulties, were dumping in the British market. By 1931 and 1932 home production of crude steel was not greatly in excess of half the 1929

[1] *ICTR* 13 Sept. 1929, p. 391.
[2] PEP, *Report on The British Iron and Steel Industry*, July 1933.

TABLE 14.1. *British production of crude and finished steel, 1929–1932* ('000 tons)

	1929	1930	1931	1932
Crude steel	9,636	7,326	5,203	5,261
Girders, joists, beams	415	374	335	n.a.
Plates over ⅛ in.	1,360	1,041	531	492
Rails, sleepers, etc.	750	601	509	373
Sheets	595	445	407	405
Galvanized sheet	843	580	447	359
Tin, terne, and blackplate	880	814	717	745
Wire rods	248	233	223	312

Source: PEP *Report on the British Iron and Steel Industry*, July 1933.

figure. Although production of sections remained fairly high, the plate trade was much more heavily distressed (see Table 14.1).

In these circumstances the closure of plants and the failure of firms went on apace. The North-East Coast, as one of the two main traditional heavy product areas, suffered severely. National output of crude steel in 1932 was 54.6 per cent of the 1929 figure; in the North-East the ratio was only 45.3 per cent. In 1921 blast furnaces, steelworks, and rolling mills in the region employed 46,000; between 1923 and 1930 the average was 39,000; and now the figure dropped still more dramatically (Board of Trade 1932: 11). Closure of works was already underway. The Newburn steelworks and mills, located on the Tyne less than a dozen miles from Consett, failed in 1924; they were being dismantled in 1929. During the twenties Palmers' great establishment at Jarrow was collapsing, and a receiver was appointed in 1933.

As foreign steel flooded in, so agitation increased for tariff protection behind which the industry could shelter while modernizing. The Bank of England, the Securities Management Trust, and the Bankers Industrial Development Company were prominent supporters of the rationalization process. In February 1932 Robert Hilton of United Steel Companies recommended to the Heavy Steels Committee of the National Federation of Iron and Steel Manufacturers that there should be co-ordination between firms on capital spending, the closing of redundant works, central selling, and the purchase of ores, scrap, etc. Although E. J. George was prominent in the industry's call for protection, the Consett directors approached such suggestions of wider co-operation with caution. They agreed that they should be represented on the proposed subcommittee to look into Hilton's ideas, but without committing the company in any way. However, the very process of change at the time made liaison between companies inevitable. From mid-1932 the imposition of an *ad valorem* tariff of $33\frac{1}{3}$ per cent was accompanied by the establishment of an Import Duties Advisory Committee, which became something of a watchdog on the industry for government. Not without difficulty the industry itself produced a more effective organization to help modernization along. The new central body, the British Iron and Steel Federation, came into existence in 1934. In February 1936 BISF decided that the chairman and managing

The Thirties

director of member firms should 'be invited to consult the Chairman of the Executive Committee as soon as the possibility of expanding their plant is under consideration'.[3] As a result of these developments, the operating milieu of the industry was transformed. Gradually the impact was felt within the regions. In this new setting of depression, nationally inspired rationalization, and merger, there was a danger that Consett would be left out in the cold and that therefore its longer-term survival might again be at risk.

In late summer 1929 Bolckow Vaughan and Co., once the leading concern in the region, was absorbed by Dorman Long. For the next four years attempts were made to extend the merged group to include the South Durham/Cargo Fleet firms, the remaining large operations in the Teesside area. Eventually these attempts failed, but the rationalization of production which they anticipated was in part carried out within the still independent concerns. The prize to be won seemed to some to justify any short-term ill effects. As one review of the Dorman/Bolckow merger stressed, it was important not to be put off by regrettable social costs: 'the stoppage of plant naturally would cause a temporary rise in unemployment in the localities concerned, but embarrassments of this nature must not be allowed to obscure the eventual and lasting benefits which will accrue from greater efficiency and reduced costs of production.' In a complex of plants such as that on Teesside, major gains might clearly be achieved from a rationalization which involved concentration as well as modernization, but if merger and reconstruction also included remote, single plant locations, there was a danger of wholesale loss. This was seen to be a possibility if Consett was brought in.[4]

Sometime in spring 1930, at a meeting of the Iron and Steel Institute, E. J. George had a conversation with Benjamin Talbot of South Durham about the proposed link of the latter with Dorman Long. Talbot's opinion was that negotiations could only succeed if Dormans took over the whole of their debentures and also paid £1.5 million in cash. They then went on to discuss an alternative, a possible amalgamation of South Durham, Cargo Fleet, Skinningrove, and Consett. The line taken by George was reported to the Consett Board as being 'to envisage some consideration of working arrangements rather than a probable physical amalgamation'.[5] This espousal of co-operation rather than merger was to be a recurrent theme. In mid-April George had discussions with Charles Bruce-Gardner, the Bank of England's adviser on steel and head of the Bankers Industrial Development Company, and also with Charles Mitchell of Dorman Long, on the matter of regional amalgamation. Mitchell suggested that, whether or not merger was possible, they might make some working arrangement over the rolling of sections. Following this, George had further talks on the same subject with A. N. McQuistan of South Durham/Cargo Fleet and James Henderson of Frodingham Iron and Steel,

[3] Minutes of Council of the British Iron and Steel Federation, 15 Feb. 1936.
[4] *Econ.* 7 Sept. 1929, p. 444; *ICTR* 18 Oct. 1929, p. 591; Boswell 1983: *passim.*
[5] CDM 6 May 1930.

Scunthorpe. Consett's Finance Committee decided to continue its interested but cautious approach. They would await further developments on Teesside, but 'if the matter should be raised with the Consett company, it would be willing to attend any meeting for discussions.' On 29 August 1930 the Consett Chairman, Mr Clarence Smith, met Arthur Dorman and Charles Mitchell. It was now revealed that Dorman Long had approached Bruce-Gardner for a loan to build a new central coking plant, and that before he agreed Bruce-Gardner wanted to know what capital expenditure was planned at Consett. Dorman and Mitchell realized that there was no possibility of an early amalgamation with Consett, but wanted a working agreement, one of whose features would be that Consett would not go ahead alone in applying to the Bankers Industrial Development Company for its own needs, and 'it being understood that any capital expenditure in the district would be done by mutual agreement'. This time Consett's reaction was decisive: 'Mr Clarence Smith had replied that he could not imagine his Board agreeing to anything of that kind', though again it was stressed that they would consider working arrangements. However, when Consett directors met four days later they concluded that the present time was not an opportune one even for working arrangements. Their reasons were that they believed they had gained an advantage because of the Middlesbrough area amalgamations—presumably due to the disruption to production which rationalization and rebuilding there implied—and as a result of their own selling efforts. Strange though this opinion may seem, they recorded a conviction that Bruce-Gardner considered Consett held the key to the region's steel trade, so that 'nothing could be done without giving consideration to its position and requirements'.[6]

Consett Iron's judgement of its significance in the reorganization of the industry on the North-East Coast was by no means matched in the assessments made by others. Arthur Dorman wrote to Bruce-Gardner on 18 September to report the meeting with Clarence Smith. He and Mitchell had purposefully refrained from talking with Smith about the 'regional' aspect of things until they had finished talks with South Durham.

> We offered to keep him fully informed from time to time of our progress, and asked him to refrain from taking any action which might be ultimately detrimental to our joint interests. The points discussed were the rolling of joists, in one or two mills only, a joint selling policy having regard to the railway carriage incurred, and generally to prevent duplication of effort. Clarence Smith listened sympathetically and undertook to bring these matters before his Board, but I have not heard from him during the last fortnight.... We feel sure that a constructive commercial policy could be formed with the Consett Company, should the South Durham negotiations be successfully completed, and that this would probably lead to a physical merger in due course.[7]

In October, in the course of refusing to entertain the possibility of merger with Dorman Long, Consett is said to have maintained that it was already

[6] CFC 24 Apr. 1930, 6 May 1930, 2 Sept. 1930.
[7] A. Dorman to C. Bruce-Gardner, 18 Sept. 1930, SMT 3/107.

'completely' rationalized, and 'almost fully' modernized. However, two months later the other, more modest perspective on their regional significance was again emphasized when Bruce-Gardner, focusing on the link between Dormans and South Durham, remarked that 'rationalisation of the industry on the North East Coast can be practically effected by the unification of these two groups and the reorganisation of their operations.' He now assumed an understanding rather than a merger with Consett.[8] The lesson of these negotiations and conversations of 1930 is clear. On the one hand was a medium-sized company which, because it had been successful, thought it was essential in any scheme of regional reorganization; on the other were the company's rivals, together with interested but outside parties, which saw it as peripheral.

In spring 1932 there was an object lesson in what might happen to a small, outlying plant when its existence seemed to conflict with the longer-term interests of bigger companies. In this particular instance the rationalization was one from which Consett would benefit, but the example's applicability to their own circumstances could not be ignored. What happened was that McQuistan suggested that his firm, Dormans, and Consett should guarantee annual interest payments on the debenture stock of the Skinningrove works of Pease and Partners—amounting in total to no more than £13,500—on condition that the works be closed for five years. As the Consett minutes recorded, 'The benefit to be derived by the steel makers would be that the orders now going to Skinningrove would go to the other works on the northeast coast.'[9] In the event this proposal was not carried through, and Skinningrove was closed only for five months in 1934 for reconstruction. In mid-1934 representatives of all the integrated firms in the area, including Skinningrove, met in Middlesbrough to talk over matters of mutual interest. Two of the main topics of discussion were exchange of orders for various sizes of sections and joists, and the possibility of a co-operative venture in new mill construction. Consett approached the question of co-operation with its accustomed caution. By this time others were pushing ahead rapidly with their modernization programmes. As they did so the balance of power in the region began to move in their favour.

When the proposed Dorman Long/South Durham merger fell through, both companies embarked on reconstruction. New plant was installed, and there was greater concentration and specialization by works. The technological gap which Consett had opened up by its modernization 10 years earlier closed. This may be clearly seen in the field in which Consett had been outstanding as a pioneer in the twenties, coke-making. In mid-1933 the annual coke capacity of the ovens operated by the Consett Iron Company was 1.3 million tons, much of this in up-to-date, or, in the case of Derwent Haugh, very modern equipment. Dormans and the South Durham group combined then had a

[8] Vaizey 1974: 63; Bruce-Gardner 1930: 32.
[9] CFC 26 Apr. 1932.

coking capacity of 1.65 million tons. This was mostly in smaller units than at Consett, and only the Cargo Fleet ovens were as modern. Moreover, the plants were scattered, and some of them were still at colliery locations far from Teesside. The development plans for the merged concern provided for a large new central coking unit, and when these fell through, Dorman Long decided to go ahead alone. In 1935 they ordered 136 Simon Carves ovens for a new cokery at Cleveland works. This was designed to carbonize over 1 million tons of coal a year; at the same time Consett was extending its capacity by an amount only one-sixth as large.[10] This new south Teesside unit not only reduced coke costs, but, through the overall heat economy of the complex-wide system of oven/furnace gas flows, improved the iron, steel, and rolling mill operations as well. Dorman Long/South Durham planners in 1930 had been dealing with six separate ironmaking establishments and eight steelworks. They estimated at that time that their rationalization programme would lower average ingot steel manufacturing costs by 4.1 per cent and plate costs by 10.5 per cent.[11] Dormans now rationalized their own over-extended operations, closing Clarence works and concentrating production of plates, rails, and structurals at fewer plants. Together with the installation of new equipment, this undoubtedly cut their overall costs of production. Any margins of cost advantage which Consett had won by its earlier enterprise were in these ways narrowed, if not removed.

During the hard years of the early thirties the dramatic decline in demand created serious difficulties for Consett Iron. By the time of their 1931 Annual General Meeting no more than four ships were being built in North-East Coast yards. Overall in the depression years 24 yards in the region went out of business—10 on the Tyne, seven on Wearside, five on Teesside, and one each at Amble and Hartlepool. So desperate was their own plight that shipbuilders were reluctant to support the steel industry demands for a tariff, even though the latter had provided them with a £125,000 subsidy.[12] In true Consett tradition the collapse in markets was turned to some good effect. Clarence Smith reported to the 1931 AGM that they had used the slackness to experiment, so that now they could supply specifications until then beyond them, including special acid steel billets for wire-drawing or for spring steel.[13] Even so, by early 1932 Consett section mills were operating at only 30 per cent of capacity and the plate mills at no more than 22 per cent. From 1932 the company was being pressed by Lloyds Bank and by the Prudential Insurance Company (who were big stockholders) to reconstruct its capital. Four years later this was done, over half the total being written off.[14]

[10] *Coal Carbonisation*, various issues.
[11] Estimates of W. McLintock, 16 June 1930, SMT 3/107.
[12] British Association 1970: 272; CFC 26 Jan. 1932, 24 Nov. 1936.
[13] *The Times*, 19 June 1931.
[14] CFC 26 Jan. 1932; Vaizey 1974: 532.

TABLE 14.2. *North-Eastern iron, steel, and finished steel production, 1937* ('000 tons)

	Pig	Steel	Plate	Other heavy products	Sheet, light products	Semis
Consett Iron	306	365	179	88	12	34
Dorman Long	1,285	1,562	228	525	95	282
South Durham	57	316	161	201	74	55
Cargo Fleet	273	320				
Skinningrove	211	200	—	168	—	7
Others	297	62	17	—	54	—
TOTAL	2,429	2,825	585	982	235	378

Source: BISF 1944/5.

By the mid-1930s the economy was recovering rapidly, but problems continued. The outbreak of the Spanish civil war in July 1936 threatened Consett iron ore supplies. It also occasioned what might have marked a return, partially at least, to the patterns of procurement of 70 years before. In 1937 a survey was made of the prospects of obtaining interests in Cumberland haematite. It proved fruitless. Meanwhile Consett yet again pressed the railway company, now the London and North-Eastern, to reduce freight charges on ore from the Tyne. This time they were unsuccessful. A little later attention shifted to improvement of the handling facilities at Tyne Dock. The old jetty on the west side was recognized as unsafe and unloading was switched to the south-east quay. There steamers of up to 28-foot draught could be accommodated and 3,000 tons of ore could be unloaded in three shifts. In this respect at least Consett retained its lead. On the Tees, ore was still being handled at 11 wharves, six of them owned by Dorman Long.[15]

During the late 1930s Consett was still second only to Dorman Long in the North-East as an agglomeration of iron and steel capacity in integrated operations; as compared with Dormans it retained the advantages stemming from compact operations involving a minimum of duplication of facilities. Combined, the capacities of the two plants at Cargo Fleet and West Hartlepool also exceeded that at Consett. In plate, though in no other finished product, Consett Iron was still a major factor in the region (see Table 14.2). Even though others were now making more headway, Consett was not resting on its laurels. In spring 1936 a report on its operations was submitted by Brasserts. A few months later, following talks with S. L. Bengston of the International Construction Company, E. J. George was authorized to obtain a 'skeleton report' on 'policy, products and sites'.[16] It is not known whether the word 'sites' means that thinking at Consett at this time ever seriously envisaged relocation of any operations. If it did, this consideration soon merged into the wider issue of redevelopment at Jarrow.

[15] European Iron Ore Co. Ltd. 1938.
[16] CFC 24 Mar. 1936, 26 May 1936, 24 Sept. 1936, 23 Oct. 1936, 25 Jan. 1937, 26 Apr. 1937.

15

Consett Iron and the Redevelopment of Jarrow

IN the inter-war years industrial development and locational decisions began to be affected to a considerable extent by social considerations and political pressures. In steel the most famous example was Ebbw Vale, but there was also lively debate and controversy about such matters as the redevelopment of the Scottish industry and the Corby project. The North-East had the problem of Jarrow, where the proposed solutions involved a major new steel plant. Like other steelmakers in the region Consett was drawn into the discussions; unlike the others it became committed to and actively participated in the redevelopment.

A model for the nationally fought battle for the revival of Jarrow had been provided nearer to Consett, at Newburn. The Spencer steelworks there had twelve 40-ton open hearth furnaces, and boiler and ship plate mills. It had been described, perhaps too generously, as 'of the most modern kind'. In the twenties the company failed and by 1927 the works was in the hands of the demolition experts T. W. Ward. In 1928 a mass meeting of Newburn employees issued an appeal to financiers for £50,000 for the purchase of the works and another £25,000 for working capital. If an adequate response was forthcoming, the men on their part committed themselves to invest 5 per cent of their weekly earnings in the business.[1] This rescue bid failed. 1929 was a good year, but ended with the onset of the great depression. Following the boost given by the steel tariff, 1933 brought a revival of business which strengthened in 1934. The upward movement in demand brought new hope, so that by this time redevelopment was in the air. It was then that the depths of the crisis at Jarrow came to the forefront of public attention and became important in steel industry discussions.

For over 80 years Palmers had been the mainstay of Jarrow's economy, and almost as dominant there as Consett Iron in north-west Durham. Uniquely, they combined iron and steelmaking and shipbuilding in one giant establishment. The iron and steel departments had eventually proved uneconomic and they were closed in the early 1920s. Except for a brief relighting of the blast furnaces they never worked again. Palmers' withdrawal from the rolling of

[1] *ICTR* 26 Aug. 1927, p. 311, 12 Oct. 1928, p. 557.

shipbuilding steels benefited Consett. Their cancellation of an order for 30,000 tons of Consett pig iron in 1921 was an immediate blow, but it was turned to good account when they not only paid £5,000 in compensation, but agreed that Consett should supply them with any future orders for ship plate.[2] During the slump the Palmers shipyard was closed, bought by the newly formed National Shipbuilders Security Ltd., and, in 1934, offered for sale on condition that it was not used again for shipbuilding for at least 40 years. Unemployment in the Jarrow area was by now at a very high level, though, because the steel plant had closed as long ago as 1921, most of the men who were without a job by 1933/4 had not been steelworkers (BISF 1941: 9). In spite of this, efforts to reduce unemployment, and passions about the issues involved, centred around a proposal for a new integrated steel plant there. Between 1934 and 1936 there was acrimonious public debate as to the wisdom of constructing a basic Bessemer steelworks at Jarrow. It was paralleled by discussion and dispute within the industry. Consett did its utmost to keep out of the limelight.

As Palmers was an important buyer of Consett steel, the latter took a keen interest in the fortunes of shipbuilding operations there. During 1932 the Consett Finance Committee received a report from the General Manager on an unsuccessful attempt by a group of Scottish steel and shipbuilding firms to bring about the closure of the Jarrow yards. In mid-1934, several months before the first steel redevelopment plan was publicly unveiled, five North-Eastern heavy steel firms met in Middlesbrough. The proceedings were reported to the Consett Finance Committee in June. Among other things, the companies 'discussed the position in regard to Messrs. Palmers Iron and Shipbuilding Company and the possibility, or desirability, of purchasing the plant for scrap purposes rather than allowing it to be bought for a song, and, possibly, operated by the purchaser'. The Finance Committee remained cautious: 'on the general question of the discussion which had taken place, the directors were in agreement, but expressed the view that the Consett company should not, without very careful consideration, take any part in the purchase of Messrs. Palmers for the purpose indicated because of the repercussive effects which such an action might have in the way of newspaper propaganda, etc, etc.' A few weeks later, after further discussion, the North-East Coast Steelmakers decided not to take steps to purchase any of the Jarrow works. By December they were discussing press reports of a prospect of reopening there. A wide, socially motivated groundswell of public opinion was now actively supporting business interests working to that end.[3]

In the Depression and in the slow recovery from it the north of Britain had extremely high unemployment levels compared with the south. In 1934 the monthly average for Great Britain was 17.3 per cent; in County Durham it was

[2] Vaizey 1974: 31, 32; CDM 1 Mar. 1932.
[3] CFC 26 Feb. 1932, 25 June 1934, 20 July 1934, 21 Dec. 1934.

34.2 per cent. Support began to come from the more affluent areas of the nation for industrial reconstruction in the ill-favoured districts. This put the investment decisions of private companies in the spotlight of public scrutiny. The British Iron and Steel Federation was pressed to support the project for a new, integrated works at Jarrow, and its North-East Coast members were necessarily brought into the forefront of the discussions. In March 1935 Consett Iron was party to an agreement to commission Brasserts to study the prospects of such a plant. It has been alleged that the involvement of the Teesside firms in this survey was a means whereby they delayed any action until they had made their own extensions, at which point the Jarrow project lost its *raison d'être*. The tone of the Consett minute hints that their agreement with Sir Andrew Duncan of the Federation to take part was also to some extent at least a delaying action:

Palmers Iron and shipbuilding Company, Jarrow—the General Manager reported the conversation which had taken place with Sir Andrew Duncan on this matter, when the North East Coast makers had arranged, in view of the surplus capacity on the north east coast, that Mr Brassert should make a report to Sir Andrew Duncan on this point for use by him, the Consett Iron Company contributing its share of the cost of this report.

By June 1936 Consett's Finance Committee was discussing Jarrow in terms which undoubtedly pointed to a determination that the site there should not be redeveloped by an independent concern. Correspondence between the new Consett Chairman, E. J. George, Arnold Foster of the Jarrow development scheme, and Sir Andrew Duncan was submitted to the meeting and discussed.

Consideration was given to the position which had arisen due to a guarantee alleged to have been given by Sir Andrew Duncan on behalf of the North East Coast Steelmakers to Mr Arnold Foster that, should the Jarrow scheme not go on, he would be refunded by the North East Steelmakers any balance of the loss of the amount of £26,500 which he had incurred on the sale of the site, and it was decided that, provided (a) the site was sold for other purposes than iron and steel works, and (b) that the other two north east steelmakers (viz. Messrs Dorman Long and Co. Ltd. and Messrs. South Durham and Cargo Fleet Iron Co. Ltd.) agreed to take their share of such a payment, Consett Iron Co. would fall into line.[4]

Later that month the BISF rejected the idea of an integrated plant at Jarrow. Paradoxically, it seems to have been at this point that Consett Iron's interest in being involved in some form of steel operation there began to grow.

The project for a Jarrow works with 0.5 million tons ironmaking capacity and capable of 0.35 to 0.40 million tons of basic Bessemer steel was killed by opposition from established North-Eastern firms. Evidence suggests that it would have been a competitive operation (see Table 15.1). Consett was reported as having been willing to join in a works of half this size, though

[4] CFC 25 Mar. 1935, 26 May 1936, 26 June 1936.

TABLE 15.1. *Estimated attainable costs for iron and steel made on the North-East Coast and elsewhere, 1935* (per ton)

	Basic pig iron		Basic Bessemer ingots		Open hearth ingots	
	s.	d.	s.	d.	s.	d.
Tyneside, new integrated works	48	6	59	6	68	0
Teesside, current	65–8		—		85–90	
Teesside, existing plant modernized	n.a.		—		71	
Teesside, new integrated works	50		60	6	67	6
Lincolnshire	42		—		59	
Northamptonshire	38	6	51	6	—	

Note: where figures are not available, this is indicated by 'n.a.'; where they are not applicable, by a dash.
Source: Tolliday 1987: 324.

Brasserts had expressed doubts of the viability of a plant so small (Carr and Taplin 1962: 534). Even after this it seems clear that Consett did not rule out the idea of involvement in a fully integrated Jarrow works, though eventually the project was to diminish into a non-integrated operation. In January 1937 a visit was made to Consett by Lord Portal, who had already been involved in investigation of the collapse of the basic industries of South Wales, and by Nigel Campbell of the Nuffield Trust Fund. A few weeks later Clive Cookson represented Consett in further talks with Portal in London. This was followed by a letter written on 23 February in which Consett Iron offered to help in the establishment of what was still envisaged as an integrated Jarrow works. However, as in their reaction to the idea of regional co-operation, they were still proceeding cautiously. The letter read:

Dear Lord Portal,

Jarrow: Proposed Iron and Steel Works and Mills.

Referring to our conversation, a full discussion took place at our meeting today. The position of Consett is still unchanged, i.e., that we are willing to cooperate with any well conceived, economical scheme of which our Board could approve. In view however of the present situation, we are now prepared to actively assist in investigating with approved expert assistance and preparing a scheme for an iron and steel works at Jarrow, on the understanding that we would not participate unless such a scheme could be reasonably considered by us, or proved to our satisfaction to be economically and commercially sound, and this involves that the following points should be covered by you to the satisfaction of our Board. (a) Reasonable security as to the supplies, source, cost and period of contract for Basic Ore, Coal, etc. (b) Cost and suitability of the site. (c) Type and cost of plant. (d) A satisfactory scheme of providing the total capital (including working capital). (e) The provision of a Quay on reasonable terms including maintenance and dredging. (f) A scheme for disposal of slag. (g) That there shall be an Agreement that the Company will not be called upon to pay any Local Rates during construction and that these charges thereafter should be on the lowest comparable basis as to assessment and amount per pound. (h) The Consett Company will be willing to be responsible for the management of the new Company on terms to be agreed. (i) The cost of any technical or commercial reports on these and other matters not to be borne by us. There may be many points not covered above but I feel certain that unless conditions (a) to (g) of this letter can be covered, neither Consett nor anyone else should embark on this project.

Yours faithfully,
[Signed] C. Cookson.[5]

In March there were discussions with Charles Bruce-Gardner, whose opinion was that there was a need for 'something in the way of a steel plant at Jarrow', and that it would be well placed for the export trade. At this point a smaller scheme than that outlined to Portal was put forward as a possible alternative.[6] There was soon further retraction on the part of Consett.

In April 1937 Consett Iron received a report from S. L. Bengston outlining a scheme for a complete, large, and, it is to be assumed, fully integrated Jarrow works. The capital outlay was estimated at £6.5 million. However, in the same month the company took a decision which seems strange if they were indeed considering ironmaking there. They were offered and declined the opportunity to take the 30 per cent Palmer share in the colliery and coke firm of John Bowes, which in the previous year had decided to build a new by-product coke plant costing £250,000 on Monkton Fell, only 2 miles from the Jarrow site.[7] When E. J. George reported to his Finance Committee on his conversations with Bruce-Gardner it was clear that a much more modest development was being contemplated, at least in the first instance. Costs were not expected to exceed a total of £1 million, of which £0.75 million would be for plant and the rest for working capital. In early summer alternative sites were considered, including 24.5 acres at Hebburn, but then, in July, following further discussions with the Treasury and with Bruce-Gardner, an agreement was drawn up for a £1 million investment at Jarrow. The Commissioner for Special Areas was to acquire the 53-acre Palmers works site, and to make it available rent free for five years. The Nuffield Trust would provide £300,000, the Bankers Industrial Development Company and Consett Iron £100,000 each, and the remaining £500,000 of the capital would be subscribed as required by the same bodies and in the same proportions. Bruce-Gardner was to be Chairman of the new concern.[8]

Even now there were hitches. For a time Wards, the demolition contractors, seemed ready to refuse to sell the site. Consett not only expressed their impatience with this, but also demanded that if they had to clear the site themselves they should be reimbursed by the Commissioner for Special Areas.

[5] CFC 23 Feb. 1937.
[6] CFC 23 Mar. 1937.
[7] *The Times*, 6 Jan. 1936; CFC 24 Mar. 1936, 26 Apr. 1937.
[8] CFC 26 Apr. 1937; *The Times*, 17 June 1938, p. 25.

Delays continued, and in November 1937, writing to Bruce-Gardner, Clive Cookson revealed their general frustration and, by implication, compared his own firm's attitude with that of the other North-Eastern steel firms only a year before. 'We at Consett have put ourselves, all along, unreservedly in every way, at the disposal of the parties in London. We have also dealt promptly with all the correspondence and documents as they have arisen. Frankly, we feel despondent about the prolonged, preliminary proceedings to which there seems to be no finality.'[9] Not until the following spring was there impressive, material headway.

In April 1938 a tender was received for a Morgan semi-continuous 12-inch strip and merchant bar mill; the order was placed in midsummer. On July 6 the New Jarrow Steel Company was incorporated as the result of an agreement between Consett Iron Company on the one hand and, on the other, the Treasury, the Bankers Industrial Development Corporation, The Nuffield Trust for Special Areas, and Charles Bruce-Gardner of the Securities Management Trust. Its articles of incorporation left a wide field open for its future enterprise, for the objects were defined as 'To carry on the trades or businesses of iron masters and founders, steel makers, steel and iron converters, constructional and general engineers, tank and bridge builders ...', and so on. New Jarrow Steel now acquired, leased, or obtained options on 95 acres of the Palmers works. It was confirmed that the new company would honour Palmers' agreement of December 1933 with National Shipbuilders Security Ltd., that no ships would be built there for at least 40 years. More important, as was shown by the mill equipment ordered, operations were to be confined to a narrow range. Clive Cookson told the Consett AGM that the money available was 'not more than sufficient for a good rolling mill' and to form a reserve for possible future installation of electric furnaces. In fact this reserve was never to be spent on steelmaking at Jarrow.[10]

For an outlay of no more than £200,000 from its own resources, Consett Iron had diversified its product range and secured an important tied outlet for semi-finished steel from its mills. The Morgan mill began rolling on 15 May 1940. In the first full financial year operations there lost £47,376 after paying depreciation; after that Jarrow earned substantial profits. During 1948 Consett Iron exercised its option to acquire the balance of the share capital of the new concern, whose works were then renamed the Consett Jarrow Rolling Mills. Eventually rolling of sections was moved there from Consett. Even though in its final form the Jarrow operation was only in the most shadowy sense a case of Consett Iron at last coming down to tidewater, it was to play a most important part in shaping postwar development by influencing the specializations of the company's main rolling mills.

[9] CFC 25 June 1937, 24 Sept. 1937, 26 Nov. 1937.
[10] *The Times*, 17 June 1938, p. 25.

16

Consett Works after World War II

IN 1937, a year of marked revival from depression, Consett's steel production of 365,000 tons ranked it seventh among the 20 integrated steelworks in Britain. Within the North-East it was second only to the Dorman Long complex, which was more than four times as big and by far the largest integrated operation in Britain. South Durham Steel and Iron at West Hartlepool, and Cargo Fleet, though closely linked in ownership, produced 316,000 and 320,000 tons respectively. During the thirties the fortunes of the South Durham group had fallen so low that it is said that at one time Consett had the opportunity to buy it for £1 million, but lacked the money to do so.

On the eve of war, Consett was busy remodelling its works. At the parent plant, weekly ingot capacity was 8,000 tons. From 1938 the company had been investing heavily in bright steel bar capacity at Jarrow. The work-force was 8,000. In the early stages of the war a new blast furnace was commissioned; in 1943 a sinter plant came into operation. Inevitably the circumstances of the time disrupted the normal ordering of business. Overseas markets were lost, new lines of production had to be followed, and raw material supplies were altered. In 1943 imported ore supplies supported less than a sixth of the Consett production of pig iron; the rest had to be made up from home ore and sinter. Even so, Consett output held up better than that of the region as a whole (see Table 16.1).

Throughout the industry there was very little wartime construction of new capacity, in marked contrast to the situation during the Great War. However, in the last two wartime years active planning for post-war expansion was underway. Consett Iron, like the other companies, began work on schemes for submission to the central planners of the British Iron and Steel Federation. Within that body discussion focused on two questions: was Consett any longer a cost-competitive location for bulk steelmaking? If so, in what should it specialize?.

There was a wide presumption among those concerned with the future of the industry that Consett's inland location must mean that costs for ironmaking were high. This was echoed in the so-called Franks Report completed in February 1945. The evidence for this conclusion is however scanty, though circumstantially the expectation seems reasonable. One indicator in its favour—though an unsubstantiated one—is that it is said that while waiting for approval of their reconstruction plans in autumn 1946,

TABLE 16.1. *Crude steel production of Consett Iron and of other North-East Coast firms, 1937, 1943, 1945, and 1950 ('000 tons)*

	1937	1943	1945	1950
Consett	365	388	382	548
Dorman Long	1,562	1,313	1,184	1,721
South Durham	316	292	305	457
Cargo Fleet	320	227	213	323
Skinningrove	200	169	161	208
Others	62	80	36	17
Total for North-East Coast	2,825	2,469	2,281	3,274

Sources: BISF 1944/5; BISF 1945; and British Steel EMRRC BISF SEC/2/1C.

Consett considered moving to a site on the Wear estuary. However, the capital cost of developing a completely new site was dauntingly high: in that year South Durham reckoned it could expand its ingot steel capacity by one-quarter at a capital outlay of £9.75 a ton, whereas at the same time the cost of greenfield construction was put at £25 a ton.[1] On the other hand, there is a good deal of evidence that the Consett cost situation was not particularly unfavourable. In fact, in the course of 1944–5, Bengston, who had obtained considerable insight into Consett operations in the thirties, observed that company figures showed that, as a result of (rather than in spite of) a coalfield location, their assembly costs for ironmaking raw materials were slightly below those on Teesside.[2] Bengston therefore looked favourably on Consett proposals for modest development spending. A few months later this opinion was endorsed by the Federation.

Imported ore destined for Consett could come into either the Tyne or the Wear. Until as late as 1948 the company entertained plans to develop new ore terminal facilities at Derwent Haugh to handle vessels of up to 12,000 tons. Such a project would have replaced a land haul on ore of about 25 miles with one of no more than 10, though at the price of sacrificing the possibility of later being able to handle ore ships of bigger size.[3] The idea was rejected by the Dock and Inland Waterways Executive. After this, with co-operation from the British Transport Commission and the Tyne Improvement Commission, Consett Iron built new ore unloading plant at Tyne Dock, a terminal able to handle vessels twice as big as the upper limit for Derwent Haugh. The ore was sent on in 56-ton wagons, trainloads of which were leaving Tyne Dock at 2-hourly intervals by 1954. Because of the steep gradients involved on part of the route, the trains were then limited to only 8 wagons. At that time small quantities of home ore from near Corby were delivered to Consett by night train, but for all practical purposes the works, as throughout most of its

[1] BISF 1946; Vaizey 1974: 104, 112.
[2] *Times Review of Industry*, Jan. 1951; BISF 1944/5.
[3] BISF Development Committee, June 1948.

PLATE 16.1. Tyne Dock ore terminal, 1957

history, depended on foreign ore. By 1956 the approach to Tyne Dock had been dredged to a depth of 35 feet; 10 ore trains now left daily[4] (see Plate 16.1).

For access to coking coal Consett was still as favourably placed as any works in Britain. Until nationalization of the coal industry on 1 January 1947, they owned seven coking-coal pits, each year raising up to 1.25 million tons of coal. All were within 12 miles of the works. By 1950 Fell Coke Works had been operating its original installation of 60 Wilputte ovens continuously since 1924. An additional battery of 54 ovens, this time of Becker design, was put to work in 1948. Fuel economy was increasing; in 1951 Consett achieved a record low coke rate of 14.7 cwt. per ton of iron.[5]

By the mid-fifties Consett was handling an incoming traffic of 62,000 tons of materials a week, and was dispatching 15,500 tons.[6] Their logistics, though by no means optimal, had now been efficiently organized. The next problem was to decide what products to specialize in, a decision in which the situations and the attitudes of the other major North-East Coast steel companies were of decisive importance.

Since the merger with Bolckow Vaughan in 1929 Dorman Long had been carrying through a rationalization of production on south Teesside. As a result Redcar now specialized in plate, Cleveland in rails, Acklam in billets and bars, while Britannia works rolled the company's heavy sections. It appears that in 1945, Consett had discussions with Dormans over a so-called 'Ten-year plan' (Vaizey 1974: 140). Nothing in particular seems to have come from these discussions, but good relations were important in view of the fact that in summer that year Ellis Hunter of Dormans became the President of the BISF and therefore a dominating influence on the final shaping of the development report for the whole industry. The position of South Durham/Cargo Fleet was in some ways even more relevant, particularly in view of the later outcome of some of the decisions now taken.

At West Hartlepool, South Durham operated a light plate mill, completed only in 1938, and had heavier plate production and some older sheet mills. There was also steel capacity and some finishing operations at Stockton. Cargo Fleet rolled rails, sections, and billets. In 1930, at the time of the merger negotiations with Dorman Long, it had been planned that heavy plate production would cease at all the South Durham works. Cargo Fleet would then give up rail manufacture and become the new group's main plant for heavy joists and sections. The separate development of the two groups through the 1930s had brought about some concentration, but had also reduced the freedom of action for post-war planning. South Durham now had plans to install a continuous strip mill at Cargo Fleet, for the extremely small capacity of 75,000 tons, a figure which caused Bengston to characterize it as 'certainly

[4] *ICTR* 1 Mar. 1954, p. 87, 24 Sept. 1956.
[5] *Chemistry and Industry*, 1 July 1950, p. 498.
[6] *ICTR* 1 Mar. 1954.

un-economical'. By 1944 BISF thinking envisaged that Dorman Long should take over the manufacture of heavier sections from both Cargo Fleet and Consett, rolling them in a new universal beam mill. It proposed that much of the West Hartlepool works should be abandoned, plate mill operations there surviving but only by working up slabs sent from Redcar. Cleveland works should take over the whole of the region's rail manufacture. Billet and sheet bar production would be concentrated at Cargo Fleet. Consett it was planned should devote the whole of an extended steel capacity of 450,000 tons to the rolling of its traditional main specialism, plate. For this a new 4-high plate mill was to be installed. It would take over the heavy plate business of the South Durham company. The planners seem to have intended that the source of supplies for the Jarrow bar mills would be switched to the new billet mills at Cargo Fleet. The Development Plan as published in December 1945 confirmed these recommendations for Consett: 'The Consett rolling mills are to be remodelled for the manufacture of medium and heavy plate and will give up the rolling of rails, billets and sections.' In efficiency and the quality of its products it seemed that Consett's new plate mill would be equalled by only one other in Britain, appropriately one to be installed by their long-term rivals, Colvilles. Such equipment would give both of them a continued viable position in the industry, although on grounds of production costs at least the advance over old-style mills seems to have been slight. (The Development Plan indicated that, taking into account their relatively high capital charges, the new plate mills would have overall costs per ton of plate of £16. 7s. 6d. as compared with £17. 0s. 0d. for the highest-cost 20 per cent of existing capacity.) More important, they would turn out a superior product. It was an expansion programme in accord both with Consett's past and with its current wishes. Unfortunately it was frustrated by the attitudes and policies of the South Durham group.[7]

South Durham had proposed to the national planners that it should rebuild the heavy plate capacity at West Hartlepool. Indeed, as BISF representatives found when they visited the works, South Durham had begun work on this new plant in 1944 without waiting for central approval. At Cargo Fleet, in addition to a strip mill, they contemplated moving into the production of beams of larger size, a scheme which they had been exploring since before the war. In the light of these plans it is not surprising that BISF's ideas provoked angry disapproval. Benjamin Talbot, who by now had been a dominating figure in South Durham's affairs for well over 30 years, put it trenchantly: 'We have continuously opposed the Federation proposals as affecting our works, which were entirely contradictory to the schemes of reconstruction previously submitted to the Federation, but which did not meet with the approval of the Federation's advisers, although we take the view that we ourselves know best the particular necessities of our plant.' South Durham argued that West

[7] BISF 1944/5; BISF 1945: 10, 20.

Hartlepool would be able to provide a wider plate than was available from other mills in the region. Therefore, as Talbot put it, 'We decline to be a willing party to depriving shipbuilders in our immediate vicinity of plate supplies of a capacity they cannot obtain elsewhere on the North East coast' (Willis 1969: 27 and *passim*). South Durham obstinacy eventually won concessions. In spring 1947, following examination by another Federation subcommittee, approval was given for the completion of the modernization of the West Hartlepool heavy plate mill. This upset plans for Consett. Unfortunately that company's management proved rather less tenacious than South Durham's in resisting an uncongenial role.

Consett Iron submitted four major alternative development schemes in 1944. All involved extensions of steel capacity, in one of the plans to as much as 500,000 tons. However, with South Durham digging in its heels on the plate issue—and also as priority in national steel development spending and in the calls made on over-extended suppliers of plant and equipment was given to the strip mill scheme in South Wales—the new Consett plate mill was deferred. This was to prove not only a long, but also a critical delay, for by the time it was eventually installed the long post-war boom in the demand for shipbuilding steels was running out of momentum. Meanwhile the Federation proposal for a Cargo Fleet billet mill was also defeated by South Durham's commitment to beam manufacture there. Plans for another new billet mill in Northamptonshire also did not go ahead. Consett's Jarrow works was a major outlet for billets, and so, with advance in its own traditional lines of business blocked (and with the Iron and Steel Act on the Statute Book by November 1949 presaging the much closer control of nationalization, and all kinds of uncertainty), the firm allowed itself to be persuaded to undertake the large-scale production of billets.

Between 1950 and 1953 a major billet mill was built at Consett. It completely altered the production schedules and the outlets for the company. In the year to September 1955, they rolled 207,000 tons of plate, but 367,000 tons of billets. In this way, Consett Iron was induced to concentrate on semi-finished steel, which, with the exception of its tied outlet at Jarrow, was for sale to rerollers and others who were located, in the main, well away from the North-East. The selling price was low in contrast to that for plate. (In 1955 the average home trade price per ton for ship plate and medium plate was £31. 19s. 0d. and £35. 8s. 6d. respectively; at that time the price for soft rerolling billets was only £26. 9s. 11d.) The warning given by Gruner and Lan 90 years before, that in a location like that of Consett it was necessary to specialize in more highly finished lines which commanded a higher price, seemed to have been forgotten (see Fig. 16.1).

Consett's directors were not particularly happy about the new specialism. The words of an unpublished company history, though restrained, reveal something of their attitude. They recorded that, having built new ironmaking plant and coke ovens and extended the steel capacity, attention had turned

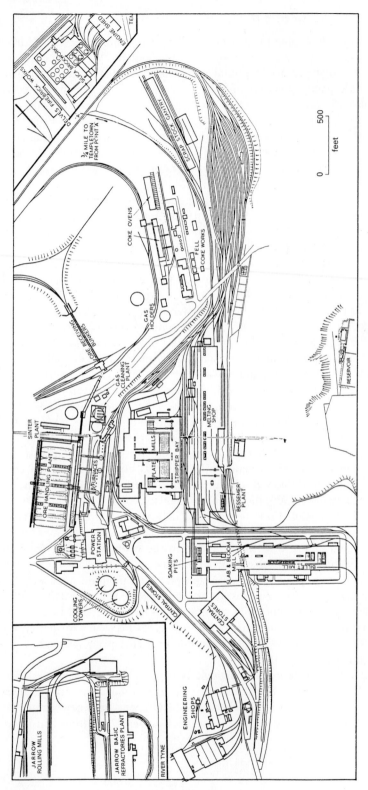

FIG. 16.1. Consett works, 1954

PLATE 16.2. Consett blast-furnace plant and part of the residential area, 1954

TABLE 16.2. *Plate production by company: North-East, 1943 and 1950; North-East and the rest of the United Kingdom, 1955* ('000 tons)

	1943	1950	1955
Consett	191	213	207
Dorman Long	212	275	333
South Durham	183	267	285
Colvilles	n.a.	n.a.	621
Appleby-Frodingham	n.a.	n.a.	400
Patent Shaft	n.a.	n.a.	59
English Steel	n.a.	n.a.	11

to the rolling mills: 'the next phase was to increase the finishing capacity of the works with a type and throughput which had to be in conformity with the overall plan for the industry, and it was considered that Consett's part in the national scheme of production and recovery was to provide additional billet and slab rolling facilities.' From once being the greatest maker of shipbuilding steels in Britain, Consett had now become a secondary factor in that trade (see Table 16.2, Fig 16.1, and Plates 16.2 and 16.3).

As well as the slabbing and billet mills, there were a number of other important expansion schemes at Consett in the late forties and early fifties. A new blast furnace was completed in 1947, and another in 1950. By the late fifties the three furnaces would be capable of making 21,000 tons of iron weekly. The open hearth furnaces were rebuilt to a larger capacity. In 1954 major extensions were made in the sinter plant. In short the commitment to expansion at Consett was not in doubt, but the form to be taken by any further rounds of increase in finishing capacity was still uncertain.

As early as 1950, with construction of the billet mill still at an early stage, the Consett Board were already looking ahead to the time when, having completed this project, they might reactivate plans to modernize their plate-making capacity. Meanwhile outsiders became accustomed to thinking of Consett as a relatively minor factor in the plate trade. After the short and not very significant first period of nationalization of steel, the task of transferring the industry back to private ownership was given to the Iron and Steel Holding and Realisation Agency (ISHRA). This body offered Consett Iron Company for sale in December 1955. A prospectus described the works: the coke ovens, the blast furnaces, the steel plant, and the slabbing, blooming, and continuous billet mill, and then added in the next sentence, almost it seemed as an afterthought, 'there are also two plate rolling mills'. Over two years later, in his annual review of operations, Clive Cookson noted that the only part of the works which was not new was the plate mills, which were too small by contemporary standards and had been practically written off in the company's books. By 1954 a new plate mill was being planned. The situation

PLATE 16.3. Consett steel plant and part of the town, 1954

was, however, far from straightforward. Consett Iron had now become an important factor in the semi-finished product trade, and therefore lacked the iron and steelmaking capacity to support both that business and extended plate production, while demand in each was booming and unsatisfied. When at last they were free to go into expansion of plate capacity they also had to install more steel capacity. By that time not only had demand begun to falter, but others too had schemes for new plate mills.

17
The New Plate Mill

IN the early part of the second half of the 1950s there took place the last major round of expansion in the British steel industry before the apparently endless upward spiral of demand faltered in 1958. At this time something occurred which in some ways was similar to the experience after the Great War—a major extension of national capacity to make shipbuilding steels, which came into production as the boom passed. In that earlier period Consett had weathered the storm as well as or better than most of its competitors; this time its success was less marked.

In 1957, helped by the introduction of duplexing—the pre-refining of the metal in a Bessemer converter, followed by finishing in the open hearth furnace—output of steel at Consett reached a record level of 966,000 tons, three times the 1939 tonnage. The company made 350,000 tons of plate, and 500,000 tons of semi-finished steel in the form of slabs, blooms, and billets. About half the semi-finished steel was sold to rerollers, largely in Sheffield, though Johnson and Nephew in Manchester were also large purchasers; the remainder went to Jarrow. (In 1957 the Consett Jarrow mills turned out almost 140,000 tons of finished products—70,000 tons of bars and sections, largely for use in the construction and engineering industries, and 66,000 tons of strip, almost all consumed by the motor industry.) At this time it is said that some of the Consett directors would have liked to enter on a bigger scale into the further processing of their own semis, though what this would have implied in the way of new products is unknown. (See Fig. 17.1 and Plate 17.1.)

Apart from an understandable wish to get back on a bigger scale into traditional lines in order to satisfy old customers, there was another persuasive argument in favour of expansion in plate: its greater profitability. By December 1957 the home trade price for 4-inch rerolling billets was £34. 4s. as compared with plate prices ranging from £42. 15s. for basic quality to £46. 15s. for shipbuilding plate. The differences were increasing—in 1955 ship plate was 20.6 per cent more expensive than rerolling billets; by December 1957, 36.9 per cent more. Although more costly rolling and finishing was involved, the margins of profitability could be widened. By 1958 about 30 per cent of the small Consett output of plate was delivered to shipbuilders (a proportion much smaller than in the past), and about 40 per cent of the total tonnage was sold within the North-East. Consett Iron was now preparing to

PLATE 17.1. Consett works panorama, 1958

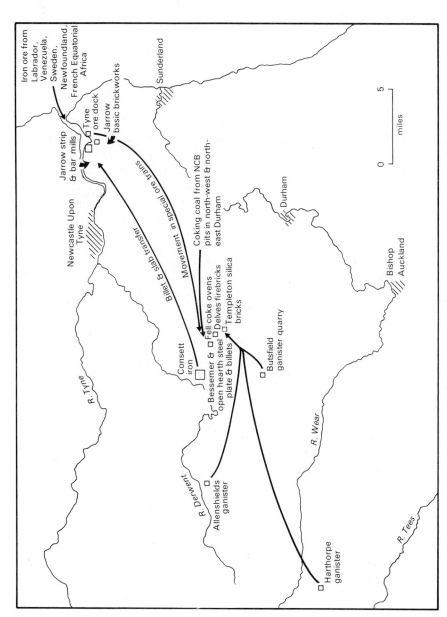

FIG. 17.1. Aspects of the operations of the Consett Iron Company in the late 1950s

embark on the next round of its expansion, centred on a new plate mill, though also entailing extension on the steel supply side.[1]

Lord Ridley announced to his Annual General Meeting that they were being urged by the Iron and Steel Board and by the BISF to increase their steelmaking capacity from 20,000 to 30,000 tons a week. In response the company planned a two-stage development. The first stage would involve a new 'oxygen Bessemer plant' and a plate mill. It would be followed later by spending on coke ovens, a sinter plant, and a 27-foot diameter blast furnace. Unfortunately almost immediately after this announcement things began to go wrong, interfering both with the smooth implementation of the programme and with Consett's capacity to support it. By the middle of 1959, billet and rerolled product markets were slipping. Nationally too the production of plate had fallen, though Consett managed to operate at 80 per cent of capacity in their out-of-date plate mills when their newer bar and strip mills were down to 60 per cent. It was pointed out that, even so, profit margins on plate were less than they would have been with a new mill.[2] The new Consett plate mill which had been looked forward to with such optimism was to take shape in an increasingly difficult business setting of which the 1958/9 recession was the harbinger.

In its Development Report published in July 1957, the Iron and Steel Board observed: 'At the present time plate is scarcer than any other steel product, and with increased requirements for ships, general engineering and pipelines, a large increase is expected in total demand' (Iron and Steel Board 1957: 54). Between 1954 and 1957 alone, plate consumption in the United Kingdom had risen by 500,000 tons or 23 per cent. The Board now reckoned that over the next four years production by the plate mills (that is, not counting the supplementary supplies of thinner plate gauges from the hot strip mills) should increase by 700,000 tons to a total of 3,030,000 tons. This expansion seemed to provide the opportunity for completing the reconstruction of the British plate capacity. The Iron and Steel Board invited firms to submit their proposals.

As with the new mills included in the 1945 scheme for the industry, the plate capacity now contemplated was to be in the form of 4-high mills, capable of turning out a product of more consistent gauge and finish than the old-style mills. It was anticipated that eventually seven of the new mills would be built. One was to be at the Black Country works of the Patent Shaft and Axletree Company, which, though linked with the shipbuilders Cammell Laird, was a comparatively small firm. A large capacity mill was to be erected by Appleby–Frodingham at Scunthorpe, and Colvilles were to build two mills, at Dalzell and at Clydebridge. Within the North-East, Dorman Long planned a universal plate mill at Lackenby, and 'may later undertake further plate

---

[1] Iron and Steel Board 1957; Campbell 1958.
[2] *Financial Times*, 10 Aug. 1959.

modernisation'. As before the most direct threat to Consett came from South Durham Steel and Iron.

As early as November 1956 South Durham had proposed a doubling of the plate capacity of the West Hartlepool works to 800,000 tons. The Iron and Steel Board now encouraged them to proceed with this scheme as quickly as possible. Two months after South Durham's proposals were submitted to the Board, Consett Iron sent in a plan for a half million ton increase in steelmaking capacity and expansion of their plate tonnage from 250,000 to 600,000 tons. Billet production they intended should increase only slightly. (Late in 1958 Consett modified this to an even smaller extension in billets.) The Board expressed regrets about the Consett billet/plate ratio, doubting the wisdom of too much dependence on plate. They also revealed reservations about the Consett location. However, in order to avoid the possibility of Consett being eliminated altogether from the plate trade, they approved the company's plans. Much more important than any dispute about the precise balance of plate and billet tonnage was the ensuing clash between Consett and South Durham.

By summer 1957 there seemed a threat both that there would be overexpansion in plate in the North-East, and that in financing their own schemes for this expansion the two companies would run out of resources. The Iron and Steel Board consulted the Treasury on the matter, and the banks, the Bank of England, and the Finance Corporation for Industry were all involved in considering possible sources of finance. South Durham was said to require £60 million or more over the next five years. After allowing for what they could themselves provide from profits, reserves, and so on they were reckoned to need from outside sources some £35 to £40 million. FCI would give a maximum of £15 million. Consett capital needs were only £44 million, of which they could raise £18 million from their own resources. In July it seemed likely that the remainder might be made up of £6 million from FCI, £12 million from the company's bankers, and £8 million by an issue of debentures. However, at this critical juncture, when much still seemed uncertain, the quality of the top management at Consett began to fall under suspicion. This is highlighted by a Bank of England memorandum of a discussion with the firm which, a year earlier, had handled the sale of Consett Iron on behalf of the Iron and Steel Holding and Realisation Agency. It read:

Sir Charles Hambro reported today that further examination of the Consett scheme only confirmed how vague the figures were. Since the death of Boot it is possible that the management of the company has not been nearly strong enough at the top. Instead of the original guess of £28 million, of which £18 million from bankers, it is conceivable that the final figure may be as much as £40 million; if so, there may be a serious question whether the forward estimates of profits will make this supportable.[3]

A suggestion was made—by whom is not clear—that the problems both of possible over-expansion in regional plate capacity and of shortage of capital

[3] Papers on Consett Iron Company (C40/846) at Bank of England.

might be eased by a co-operative arrangement. The proposal was that South Durham would build its proposed mill forthwith, but Consett would postpone reconstruction, meanwhile improving the primary mills from which, for four or five years at least, it would send slabs for further rolling on the new West Hartlepool plate mill. As a result the latter would be assured a favourable load factor. Perhaps it is not surprising that there was no welcome for this idea at Consett Iron. Meanwhile South Durham found sources of finance which enabled it to go ahead with its full scheme. A few months later yet another new South Durham initiative transformed the regional context for development.

Benjamin Talbot, South Durham's long-serving Managing Director, and from 1940 its Chairman too, had died in 1947. He was succeeded at once as Managing Director, and three years later as Chairman as well, by his son Chetwynd Talbot. Chetwynd was a remarkable, individualistic, isolated leader of industry, who found co-operation with others difficult, and who was inordinately jealous of the position of his company (Willis 1969: 47). He was mildly enough recalled by one Consett director some years later as 'a law unto himself'; Colville company recollections confirm his idiosyncracies, which are indeed admitted even in the history of the South Durham company which was published after nationalization. Even at this late stage in the collective development of British steel Chetwynd Talbot, like his father, would not fit in with industry-wide plans unless they were also unquestionably in the interests of South Durham. In January 1958 he made a bold attempt to cut Consett out of the development programme for plate. To his Annual General Meeting that month he outlined a plan not just for a plate mill, but for a new, fully integrated iron and steelworks on a greenfield site at Greatham to the west of West Hartlepool. The new Greatham heavy plate mill would by 1962 give the company a combined output for all types of plate of 1 million tons a year. He went on with amazing audacity to ignore the central planning suggestions which had some years before been so hard on South Durham; indeed, he stood them on their heads. He said:

taking into consideration this additional plate capacity which can ultimately be secured, in my view a great opportunity now presents itself as indicated in the White Paper issued in the immediate post-war years for the concentration at one plant on the North-East coast of heavy and medium plate production. To reinforce these remarks, it would be in effect a continuation of the policy which has been accepted in other sections of the Iron and Steel Industry for the fullest usage to be made of modern plant possessing the highest level of efficiency.[4]

Before this South Durham proposal was made public, the Iron and Steel Board had already given approval to Consett Iron for the building of coke ovens, sinter plant, a new blast furnace, a melting shop, and plate mill. Consett later rephased their proposals, deferring the coke and ironmaking sections of

[4] *Manchester Guardian*, 7 Jan. 1958.

The New Plate Mill

it. Talbot opposed Consett's plans but, in spite of this, in the month following his outspoken speech to his shareholders, the Iron and Steel Board also approved the amended Consett scheme. Completion was expected earlier than it had been for the more comprehensive development—in 1961 rather than 1962. Unfortunately, plate demand was now dropping dramatically. The 1957 Iron and Steel Board report on development anticipated that 1959 plate deliveries to the national market would be 200,000 tons greater than in 1957; they were 400,000 tons less. Deliveries of finished plate from United Kingdom production in 1962 were projected to be 50 per cent up on the 1957 figure; in fact the level achieved was well over 20 per cent down on that year's tonnage. By the mid-1960s the plate-making capacity of Britain had been effectively modernized (See Table 17.1) but the market had shrunk. (The relationship between shipbuilding and steel output at this time is indicated in Table 17.2.)

Although earlier estimates had envisaged an even higher figure, between 1958 and 1961 the Consett Iron Company spent some £28 million on the new plate mill and the supporting steel capacity. It raised £10 million of the total internally, and borrowed £9 million from the Finance Corporation for Industry and the rest from Lloyds and Barclays banks (Vaizey 1974: 177). In contrast to South Durham, which installed open hearth furnaces in its wholly new Greatham works (making it, so it is said, the last important installation in Europe of that soon-to-be-superseded technology), Consett Iron opted for the new oxygen steel process. The new rolling mill, named the Hownsgill Plate Mill, after the deep valley which lies on that side of Consett, was commissioned in September 1960. It had a rated capacity of 10,000 tons a week. Despite the fact that its situation some way apart from the rest of the works increased internal handling costs, it was an efficient unit, and when it reached an 80 per cent operating rate proved very profitable. (As with the new mills which had been installed in the twenties, it became profitable at the start partly by eating its way into some of the markets which South Durham could not hold on to before its own new works was completed.) Even so, in the early sixties acute recession of trade coincided with a period of maximum capital outlay, and the Consett financial situation was sometimes very finely balanced.

In 1961, reviewing the plate market, the Iron and Steel Board recognized that 'The progressive increase expected from 1957 to 1962 has not materialised, principally because of lower activity in shipbuilding' (1961: 37). In the following year the North-East, because of its marked emphasis on heavy steels, had the lowest utilization rate for steel capacity among the 10 steel districts of the country. In 1963 it had the second lowest. However, demand for plate was again buoyant in 1964 and the region's operating rate at 93 per cent was then well above the national average. Because of the new mill, Consett Iron had increased its interest in the trade and now tried to find the business to keep a high-capacity, efficient, but highly capitalized operation as well used as possible (see Table 17.3). One way of improving the situation was to open up new markets. From the mid-1950s South Durham had been

TABLE 17.1. *Plate capacity, 1955, 1960, and 1965* ('000 tons)

	Category A (first-class modern plant with effective scale economies)		Category B (older and technically dated plant, but still competitive)		Category C (old plant needing replacement)		Total plate capacity
	Tonnage	% total	Tonnage	% total	Tonnage	% total	
1955	500	20	500	20	1,250	50	2,250
1960	1,613	56	749	26	432	15	2,794
1965	3,340	90	—	—	285	8	3,625

Note: Percentages do not total 100, for some additional plate came from the strip mills.

Source: Iron and Steel Board 1964.

The New Plate Mill

TABLE 17.2. *Relationship between British shipbuilding and steel production, 1952–1969*

Year	Shipbuilding (gross tonnage launched, '000s)	Output of plate, 3mm thick and over ('000 tons)	Gross deliveries of plate from UK production ('000 tons)	Deliveries of steel to shipbuilders and marine engineers ('000 tons)
1952	1,303	2,262	n.a.	n.a.
1953	1,317	2,467	2,448	n.a.
1954	1,409	2,491	2,458	867
1955	1,474	2,452	2,434	808
1956	1,383	2,659	2,664	928
1957	1,414	2,877	2,887	978
1958	1,402	2,588	2,521	849
1959	1,373	2,340	2,288	636
1960	1,331	2,881	2,803	636
1961	1,192	2,780	2,808	679
1962	1,073	2,281	2,274	506
1963	928	2,553	2,492	544
1964	1,043	3,352	3,260	652
1965	1,073	3,172	3,128	585
1966	1,084	3,103	3,052	531
1967	1,298	2,998	2,944	418
1968	898	3,266	3,252	507
1969	1,053	3,592	3,703	574

Sources: BISF *Annual Statistical Reports* and Iron and Steel Board *Annual Reports*.

TABLE 17.3. *Output of finished products by Consett Iron, 1958 and 1961 (% of total)*

	1958	1961
Plate	25.0	33.3
Billets for Jarrow	25.0	25.6
Billets and other semis for sale	50.0	41.1

fabricating an increasing tonnage of plate into large-diameter pipe, mainly for oil and gas pipelines. To some extent following this lead, Consett now began to produce spirally welded pipe. However, this was never able to take more than a small tonnage, and does not seem to have been very successful. In 1960 an opportunity for entry to a much more expansive market seemed about to open, only to evaporate. At that time the Ford Motor Company was induced to consider Washington, County Durham, as a possible site for a new automobile plant which would be equipped with body shops. Consett let it be known that if this plan went ahead, it would build strip mill stands on the end of the plate mill and so be in a position to roll the strip sheet the press shops would need. Instead, Ford chose the site at Halewood, Merseyside. By July 1964, though demand for plate nationally was recovering, Hownsgill was still operating at

TABLE 17.4. *Consett Iron Company, production and financial record 1956–1966*

Financial year ending	Ingot production ('000 tons)	Trading surplus per ingot ton (before depreciation)			Dividend on ordinary shares (%)
		£	s.	d.	
Sept. 1956	879	5	5	0	7.5
Sept. 1957	966	5	10	3	8.75
Sept. 1958	855	4	7	2	8.75
Mar. 1959 (6 months)	334	3	18	5	4.37
Mar. 1960	904	4	6	5	8.75
Mar. 1961	979	5	2	4	9.5
Mar. 1962	856	2	11	11	2
Mar. 1963	696	2	8	0	nil
Mar. 1964	856	3	6	4	nil
Mar. 1965	1,052	5	6	2	10
Mar. 1966	992	3	13	2	5

Source: Consett Iron Company, *Annual Report*, 1966.

only about 75 per cent of capacity. It was at this time that a co-operative agreement with Dorman Long seemed to promise yet another possibility of a breakthrough to better loading. For a time it was contemplated that Dormans might buy a half share in Hownsgill, so enabling them to avoid outlay on a new sheared plate mill at Redcar. However, this approach to the problem was not followed through. Instead the two companies agreed on a co-operative rolling programme, Hownsgill producing sheared plate for Dormans, and Dorman Long's Redcar mill rolling universal plate for Consett. The existing sheared plate mill at Redcar was closed. Roland Cookson, acting Chairman and then Chairman from 1964 to 1967, assured Consett shareholders that it was 'confidently expected that these arrangements will enable each company to give improved service to customers consequent on the most efficient plant utilisation'.

In spite of the high expectations from these arrangements, the manufacture of plate did not again become as important a feature of Consett's output as had been envisaged in the mid-fifties. In 1965 plate still made up only one-third of the finished steel output. No more than 1 ton in every 3 of this went to the shipyards. Meanwhile South Durham was expanding not only plate capacity, but also the ability to further process much of it. Their own pipeworks had taken 1,000 tons of plate weekly in 1956, and took four times that tonnage a decade later. In 1966 they went so far as to plan another pipe mill at Greatham to double their plate consumption yet again to a weekly total of 8,000 tons. It was at this time, as Consett's profitability was visibly declining (see Table 17.4), that outside developments occured which shattered Hownsgill's long-term prospects. During 1966 the leading firm in the British tube and pipe business, Stewarts and Lloyds, looking for a lower-cost location than Corby, approached South Durham for co-operation on a proposed major Teesside tubeworks. By the end of that year Dorman Long had also been brought in, and a merger between the three companies was announced. The new Stewarts and Lloyds-inspired tubeworks would, it was believed, consume in the order of 250,000 tons of plate annually. As a result, before the nationalization of the industry in 1967, Consett Iron was again isolated in the region's steelmaking, though, as in the plans for Teesside consolidation 35 years earlier, there were suggestions that it might become associated with the new partners at a later date.

Though its results varied a good deal, and occasionally the old sparkle seemed to come to the surface, Consett's general financial performance was worsening and its viability was once again called into question. In January 1964, in his annual statement, Chetwynd Talbot made another scarcely veiled attack on the Consett plate mill. His own company, he pointed out, was now the largest platemaker in the United Kingdom, and its financial results showed how profitably it could operate in the trade. He continued:

The levels of profits which were secured during the latter months of the last Financial Year and have shown a further increase during the first two months of the present Financial Year, clearly indicate the great potentialities of modern integrated plants situated on tidal waters when operated at an efficient level of potential capacity. The life of plants which do not possess these advantages under conditions appertaining today, is clearly limited. In order to compete effectively in world markets under the present levels of intensive competition, not only in the direct sale of steel products, but for the benefit of the steel consuming industries, it has become essential that modern plants on tidal waters should be loaded to their full capacity and those plants not possessing these advantages, particularly in plate production, where competition has already reached severe proportions, should be encouraged to transfer their activities to ranges of products where the disadvantages of location would not represent such an insoluble handicap.

The strain, even the paranoia, under which this statement was made may be appreciated from its long-winded, repetitive character. The statement was also a reflection of the strong self-interest of South Durham, but the suggestion that Consett was not only a disadvantaged, but in practice might prove an unviable location was not one which could be lightly denied.

18

The mid-Sixties: Consett's Location Challenged and Defended

DURING the 1960s costs in British steelmaking rose steadily, with the most dramatic increases being in coking coal. Meanwhile the Iron and Steel Board kept home steel prices low. Inevitably profit margins narrowed. Yet at the same time further expensive redevelopment became desirable in order to meet the increasing competition now being experienced in markets, both at home and overseas, which were no longer expanding apparently endlessly as they had done in the fifties. In short, the ability to finance investment from earnings declined as the need for reconstruction increased. After reaching a peak in the early sixties, capital expenditure fell away.

In the mid-sixties Consett works was operating with much new plant, and still seemed a successful operation. Its 1964/5 steel output was an unprecedented 1,052,000 tons. Moreover, whereas South Durham's unit steel costs (and therefore, despite their new plant, their finished rolled steel costs as well) were inflated by continuing use of the open hearth process, Consett was now switching over to the new oxygen steel technologies. As late as July 1964, 60 per cent of Consett's steel was made in open hearth furnaces; by December 1966 the last of these had been closed down. Roland Cookson remarked in summer 1965 that, though the trials which they had made of the Kaldo oxygen steel process had shown that it 'has its value', the LD process was the right one for them. Three years later they replaced the single Kaldo converter with another LD unit. Rivals felt that Consett Iron had wasted its resources on the Kaldo process.[1]

Nationally a desire for economy in capital outlay seemed to argue strongly in favour of extension of existing plants. By 1965 the Consett management claimed to have recognized what this meant for them in terms of prospects for further growth. The Chairman explained that 2 million tons, twice their current, record, steel output, could 'in due course ... be achieved with an additional capital expenditure which would be low in relation to the sums that would be involved on a greenfield site'. To reach this target it would be necessary in the first instance to improve the existing blast furnaces so as to increase supplies of molten iron, but 'our longer term plans include additional

---

[1] T. Craig of Colvilles in conversation, Sept. 1985.

iron making capacity...'.[2] Seven years earlier they had deferred Stage 2 of their development plan which had provided for new coke ovens, sinter plant, and a 27-foot diameter blast furnace. In 1960, when Hownsgill mill was under construction, reference had been made to a possible 28-foot furnace. In fact Consett was never to be equipped with more than its existing three 20-foot 7-inch furnaces. Already, at this high point of production and of forward thinking, challenges were arising in more acute and persistent form than ever before. They now came from a wider and less obviously self-interested range of critics than South Durham Steel and Iron.

In the late fifties the Iron and Steel Board had considered the problem of what it felicitously called 'obsolescence of location'. This was essentially the reverse of the ideal situation which the Departmental Committee had analysed 40 years before. Put thus it seemed almost a direct indictment of Consett:

steel works may be affected by obsolescence of location as much as by obsolescence of plant. There are iron and steel works in unsuitable places, far from their raw materials and markets. They were put there originally on local ore fields or coal fields which are now worked out. Some have been able to adapt themselves efficiently to these changed conditions, but others have not. At an integrated works making steel from its own pig-iron, as much as six tons of materials may enter the works gate for every ton of finished steel that leaves it, and the cost of bringing in raw materials is a large part of the total cost of manufacture. An ideal steel works would be close to ore and coal supplies and near its market. If it were based on home ore it would be on the ore field itself, but if it were based on imported ore, it would be immediately beside a deep water dock. In this sense there are few ideal sites for steel works and their location has generally to be a compromise, but some sites are much closer to the ideal than others. (Iron and Steel Board 1957: 66.)

Consett clearly did not rank high in these respects. In the year in which those remarks were made it turned out 15,000 tons of iron weekly. To do so it had to assemble some 30,000 tons of iron ore, 11,000 tons of coke (requiring the movement of about 16,000 tons coal), and 5,300 tons of limestone—a total inward movement of 51,000 tons. Steel ingot production was 18,600 tons, and 16,000 tons of finished steel products were delivered from the works. In contrast to the extreme case the Iron and Steel Board postulated, the ratio of inward to outward movement was only about 3.2:1, but the situation was certainly far from ideal. In some ways it had worsened since the matter was considered by Franks.

In 1944/5 the existence of collieries working the coking coal of the area had been held to justify the continuation of a plant on the western edge of the Durham coalfield. Now, however, the fuel supply situation was less defensible. There were two main reasons for this: the continuing progress of fuel economy in iron smelting, and the decline in the coking coal production of north-west Durham. Partly because of the use of richer foreign ores, partly as a result of

[2] R. Cookson to AGM, *Guardian*, 19 July 1965.

the employment of larger and more efficient blast furnaces, and also partly as a consequence of the greatly increased consumption of sinter rather than raw ores, the amount of coke used per ton of iron was falling rapidly throughout the industry. In 1946 the coke rate nationally was 22 cwt. per ton of iron; by 1958, 17.9 cwt. (at which time the Consett rate was already as low as 14.6 cwt.). By 1961 the national figure was 16 cwt., and in 1965, 13.6 cwt. (After this the downward trend proceeded rather more slowly to 12.1 cwt. in 1976.) Such savings reduced still further the relative attractions of an inland coalfield location compared with an orientation to the ore terminal. (In fact these figures understate the case, for each ton of coke required the processing in the ovens of approximately 1.5 tons of coal, and it was this which in most instances had to be hauled to the ironworks.) At the same time as this reduction in use, the coking coal production of north-west Durham, still important at the time of nationalization, was collapsing (see Fig. 18.1). The NCB, as it found it necessary to close mines in the area, increasingly supplied the Consett ovens from collieries east of the Great North Road. Though at the end of the sixties transport costs on coal used per ton of iron were still less than half the United Kingdom average, the erosion of the works' materials assembly position was accelerated by this change.

Attention also focused yet again on the situation for delivery of imported iron ore, and the limitations of Tyne Dock became more obvious. In the thirties Consett had operated a fleet of ships of about 3,000 tons.[3] After the post-war construction of the new facilities at Tyne Dock, the company was able to handle bigger ore carriers there than could be handled at any other port in the United Kingdom until the late sixties. Up until 1958 the average carrying capacity of the vessels discharging at Tyne Dock was 15,000 tons, though the record had been a 28,000 ton cargo. Already there were indications that this might not be big enough, and that in new ore supply situations Consett would either be unable to participate or would do so only with difficulty. As early as 1957 the Iron and Steel Board's *Development in the Iron and Steel Industry: Special Report* had recognized the changing conditions of the international iron ore trade. With rapidly increasing tonnages, and longer hauls, American and Swedish companies were already operating carriers of 20,000 to 30,000 tons. Some United States interests had even ordered 80,000-ton vessels. By contrast, British developments were conservative. The Board commented, 'the policy of the United Kingdom steel industry has been very largely controlled by the size of the ports serving the steel works. At Barrow, Workington, Irlam and Port Talbot, only 8–9,000 tons carriers can be accepted, and there is only one dock at present, Tyne Dock, which can readily take a 20,000 ton carrier' (Iron and Steel Board 1957: 44). As the size of ore carriers continued to increase, its leading rank was a distinction on which the Tyne was unable to build much further. By 1960, it is true, the Consett terminal had been

---

[3] *Consett Magazine*, Mar. 1958, p. 3.

The mid-Sixties: Consett's Location

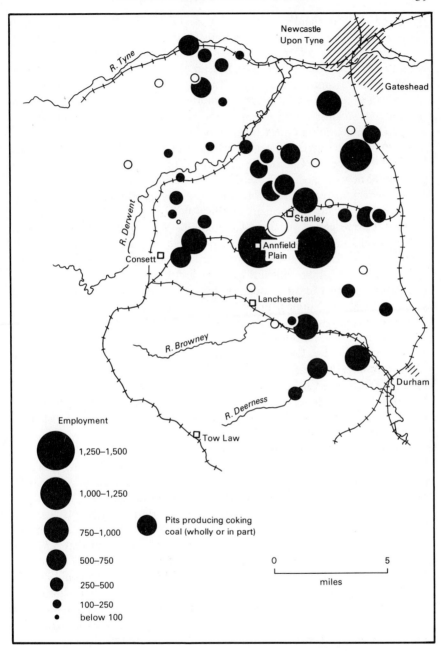

FIG. 18.1. Collieries in north-west Durham, 1960

improved so as to handle a fully laden carrier of 35,000 tons. At that time Newport and the new General Terminus Quay in Glasgow, with 23,000 tons maximum, were next in rank among 18 British iron ore ports. Three years later, though the Tyne remained in the lead, the gap was narrowing, notably as a result of developments at Dorman Long's Teesside wharf. Bigger schemes now began to come forward in greater number. The condition of the Tyne and the relative smallness of Consett's ore requirements made it difficult and ultimately impossible to meet this competition.

It began to be suggested, again harking back to the principles championed by the Departmental Committee at the end of the Great War, that Britain must build new ore docks. These should be designed to suit a new generation of ore carriers, instead of, as before, the vessels being built to fit existing docks. The penalties for not modernizing in this way were now recognized as serious. As Fred Cartwright of the Steel Company of Wales put it in an address on British steel to the Institute of Transport: 'So long as its materials and its products are carried in little ships and little wagons in small trains, then it will be just that much more difficult to survive.'[4] Consett at that time deserved castigation less than most companies, but other areas now pushed ahead. In summer 1965, the National Ports Council announced big ore dock developments for Immingham, Port Talbot, and Newport. (The plans for Newport were subsequently cancelled.) The channel of the Tees estuary was to be dredged to allow it to take vessels of up to 65,000 tons. At the same time the NPC recommended that no major expenditure or expansion should take place on the Tyne. There Consett's ability to follow the new lead was also limited by the load factor it could have provided for any enlarged ore terminal. Its relatively small size and isolation from other plants requiring imported ore were beginning to tell against it.

In 1963 626,000 tons of ore were discharged on the Tyne; notwithstanding their current deficiencies as terminals, the Tees, Port Talbot, Newport, Birkenhead, and the Clyde each handled more. Unloading facilities to handle bigger vessels would have increased the throughput capacity, but Consett was too small to fully employ such an extension.[5] In 1961, Consett management had contemplated a 1965 iron capacity of 1 million tons. Assuming that the ore was of 58 per cent iron content, the average at the time, and that it was all imported through the same terminal, this would involve handling 1.7 million tons of ore a year. Even with the existing facilities, Tyne Dock could in 1961 unload as much as 26,000 tons of ore in 48 hours. At the more modest rate of 20,000 tons in the same period, no more than 170 working days would be required to supply the ore needed by an extended plant. For the remaining almost half of the working year the terminal would be idle. In a deep-water terminal the size of the individual cargo would be much larger, and the

---

[4] Mills 1962; *S and C*, 1 June 1962, p. 1040.
[5] Iron and Steel Board 1961: 83; Iron and Steel Board 1964: 98.

economics of bulk transport required that time in port be kept to the very minimum; thus the problems of a low utilization rate for the terminal would become even more serious. Only if Consett capacity could be increased on a still greater scale could it use such an ore dock economically. It soon became clear that the rest of the industry was convinced that Consett should not increase further in size.

In March 1966, as the prospect of nationalization loomed, the British Iron and Steel Federation responded to the charge that it had not undertaken sufficient expansion by appointing a Development Co-ordinating Committee. This was under an independent chairman, Sir Henry Benson, an accountant, and included two other men from outside the industry. Its three members from the industry were the chairmen of the Steel Company of Wales and of the United Steel Companies and, from Teesside, Dorman Long's chairman, E. T. Judge. In July the committee published its recommendations as *The Steel Industry: The Stage 1 Report of the Development Co-ordinating Committee of the British Iron and Steel Federation* (BISF 1966; the implied further stages of the Benson Report were never published). Radical changes in the industry were proposed. Altogether it was anticipated that by the mid-seventies 65 per cent of existing plants would have been closed and jobs in the industry would fall from 317,000 to 215,000. Five districts were identified as possessing advantages which made them suitable as expansion points:

The major determining factor in the location of an integrated works is easy access to the deep water by which ore (and possibly coal) will be imported. Five existing iron and steel making areas in the United Kingdom should be well located for access to deep water by the early 1970s. These locations are: South Wales, North Lincolnshire, Teesside, North Wales (Deeside utilising the Mersey) and Scotland (Clyde). (BISF 1966: 62.)

The conclusion could only be that, among others, Consett should not remain a large-scale steel centre. A few months later the merger of Dorman Long, South Durham, and Stewarts and Lloyds was yet another indication of Consett's exclusion from the mainstream of advance. The Consett Iron management were not slow to recognize the implications of both these developments. They reacted positively, though in doing so they were led to make what can only be judged to have been exaggerated claims for the viability of their plant. The parallel with earlier reactions from South Durham Steel and Iron is noticeable.

In September 1966 the Consett Chairman, in the course of his annual remarks to shareholders, referred to the Benson Report. Consett had, he observed, supported the establishment of the Development Co-ordinating Committee, and they endorsed the need for further rationalization in British steelmaking.

Nevertheless we were surprised and profoundly disturbed when the first part of the Committee's report was published in the middle of July, to find that Consett was not included in the expansion plans. Indeed, though it was not specifically stated in the report, the conclusion which has been generally drawn is that Consett is marked down for closure during the 1970s. This is a conclusion we cannot for one moment accept.[6]

S. C. Pearson, the Managing Director, now contended that Consett costs of production compared favourably with the best in the country. The evidence for this was not made public and on prima-facie grounds it seems doubtful whether, if a balanced view was taken of what constituted production cost, a really convincing case could be made for such a claim. In spring 1967 it was stated that 40,000-tonners could be accepted at Tyne Dock. Thereafter the visions of Consett's defenders seemed to take wing. For example, negotiations which were said to have been underway with the Tyne Improvement Commission before publication of the Benson Report were now said to have borne fruit in a 'finalized' scheme for dealing with ships of 65,000 tons at a capital cost of £3 million. Even more recently, so another statement went, it had been shown that the Tyne could be equipped to deal with ore carriers of as much as 100,000 tons, and accordingly a revised development scheme was to be prepared.[7] Notwithstanding these spiralling figures the situation was still fragile. The contrast with Teesside was marked, and unfavourable. Dorman Long had six furnaces there, making a little more than twice as much iron as Consett, from ore which came into wharves about a mile from the furnace stockyards. For Consett to tranship a tonnage so much smaller and then to haul it 28 miles inland and up to a height of nearly 900 feet above sea level was a situation which it would prove more and more difficult to justify. There were other problems.

As with the ore terminal, so with size of steel plant, Consett's management reacted strongly to the Benson recommendations. In 1965 they had been looking ahead to a future steel capacity of perhaps as much as 2 million tons, or roughly twice the current level.[8] Benson advocated works of 3 to 3.5 million tons. Within two months of that report, spokesmen for the company were claiming that technical investigation had proved that it could be developed up to that tonnage, and at the very low capital cost of £40 million. Pointedly, the Consett Chairman remarked: 'According to our information this is a substantially lower capital investment than is involved in some of the comparable development plans visualised by the Benson Committee.'[9] After further consideration, and just before the nationalization of the industry, this plan for major expansion at Consett was reaffirmed:

The Board is satisfied after very careful consideration, that the future development and substantial extension of Consett steelworks is sound and viable. They believe that this

[6] *Guardian*, 7 Sept. 1966.
[7] Ibid., 18 Apr. 1967.
[8] Ibid., 19 July 1965.
[9] Ibid., 7 Sept. 1966.

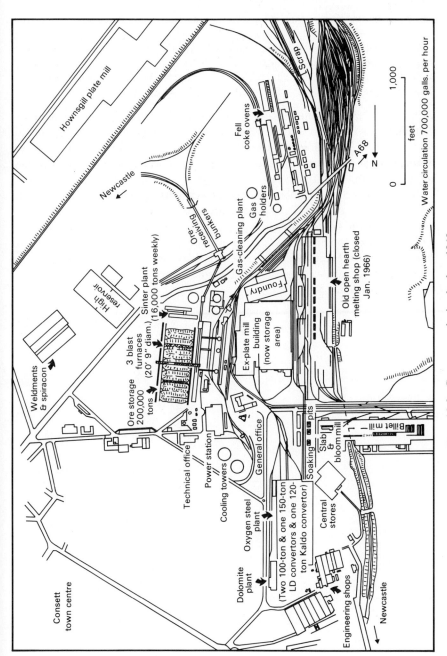

Fig. 18.2. Consett works in the late 1960s

should be undertaken within the next few years as one of the first steps in concentrating production in low cost works. This view is put forward after taking into consideration Consett's geographical position for the reception of imported ore and the large tonnages of raw materials other than ore which are required to make steel. Consett's position as a large supplier of steel to heavy and light industry makes it a most suitable growth point in the Northern Development Area.[10]

Such an expansion would obviously improve the load factor of an extended ore terminal, but the most critical consideration was whether the capacity could find a market. In fact it is impossible to avoid the conclusion that here yet again was the same sort of special pleading in which South Durham had indulged; with Consett it seems to have been even less justified.

In 1967 the major steelmaking concerns were nationalized. That year Consett had a steel production 117 per cent greater than in 1945 (see Fig. 18.2). Apart from that at the combined North and South works at West Hartlepool, this had been much the biggest expansion in the North-East. Notwithstanding the bold claims now being made on its behalf, there remained doubts and fears as to whether Consett was big enough and sufficiently well located to survive in the new situation.

[10] Ibid., 7 June 1967.

19
Consett within the British Steel Corporation

NATIONALIZATION brought a new, wider framework for decision-making, though in the early years the management of the various works found it difficult to think in these terms, and all too often fought for their corner of the ring. Three critics of BSC have divided the history of the Corporation up to 1980—the years relevant to operations at Consett—into three distinct periods. Until 1971 a 'Heritage Strategy' was followed. This was succeeded by a Ten Year Development Strategy, which, notwithstanding its name, lasted only to 1978, when it was replaced by a policy of retrenchment (Bryer, Brignall, and Maunder 1982: 73).

Consett entered the British Steel Corporation with impressive achievements over the post-war years, but with many question-marks about its future. Though the North-East Coast share of national output of crude steel had fallen from 22.98 per cent in 1951 to 21.20 per cent in 1960 and 17.05 per cent by 1967, Consett had maintained its relative position *vis-à-vis* the South Durham group, and most impressively improved it in comparison with Dorman Long. Yet its whole material supply situation was in doubt, and it stood outside the new, Teesside regional grouping. Finally, the still independent management of Consett had seemed to be ready to push ahead with expansion on a very large scale. BSC was clearly in danger of being rushed into projects which might not fit in with any long-term, Corporation-wide development strategy.

While questions of the size and shape of Consett were now to be considered in a national context, the future of the works was becoming more and more significant to the economic well-being of north-west Durham. With the rationalization and eastward shift in coal-mining, steelmaking at Consett had become the largest single source of employment. As late as 1956 there were 20,000 coal-miners in north-west Durham; by 1968 only 3,000 remained, less than half the work-force at Consett works. (The number of miners continued to fall until in the early 1980s the last pits in the district were closed.) (See Tables 19.1 and 19.2 and Fig. 19.1.)

In November 1966, the Chairman of the Steel Nationalisation Organising Committee, Lord Melchett, visited Consett, He assured local union leaders that social and other considerations would be taken into account in long-term development planning for steel. At the same time, for the company, S. C. Pearson reaffirmed both the low costs of production at Consett due to modern

TABLE 19.1. *Number of employees by sector in the district of Derwentside, 1951–1976*

	Agriculture, forestry & fishing	Mining and quarrying	Manufacturing	Services	Total
1951	662	16,874	8,745	13,458	39,739
1961	571	12,820	12,188	14,396	39,975
1971	308	2,528	13,172	13,798	29,806
1976	407	286	12,365	15,252	28,310

Sources: Durham County Council, Derwentside District Council, and Department of Employment.

TABLE 19.2. *Employment in north-west Durham and Consett, 1956*

	Total	Coal and steel (%)	Services (%)
North-west Durham	45,000	50	35
Consett Employment Exchange area	18,000	60	36

plant, and his contention that, effectively if not cartographically, Consett was nearer to tidewater than most of the Benson expansion schemes.[1] After Consett had spent over £30 million on reconstruction in the previous few years, during summer 1967—almost on the eve of nationalization—plans were submitted to the Iron and Steel Board for what was designated the second stage of the Consett development plan. This involved extension of the oxygen steel plant, improvement to existing equipment, and more rerolling facilities. The company claimed that 'in reasonable trading conditions the implementation of this scheme would increase output by 25% and profitability by 55%', all for as little as £13 million. Their spokesmen also referred, though only in very general terms, to a further third-stage expansion, to a capacity of 3 or even 3.5 million tons. It was reckoned that this could be obtained more cheaply than in their estimates of the previous year, perhaps for a further outlay of £30 million.[2]

For a time, extensions continued at Consett, even if on a more modest scale than before. In mid-1968, claiming (though not providing evidence) that Consett was the lowest-cost steelmaker in BSC, the leader of the local branch of the Iron and Steel Trades Confederation, John Page, pressed the northern group of Labour MPs to find out what BSC had in mind for the plant. The immediate auguries were favourable. That summer a new LD converter, of 165 tons as compared with the two 120-ton vessels installed four years earlier, was brought into production. Ingot production at Consett in 1966/7 was only 876,000 tons; new equipment increased the nominal capacity there to nearly 2

---

[1] *Guardian*, 30 Nov. 1966.   [2] Ibid., 7 June 1967.

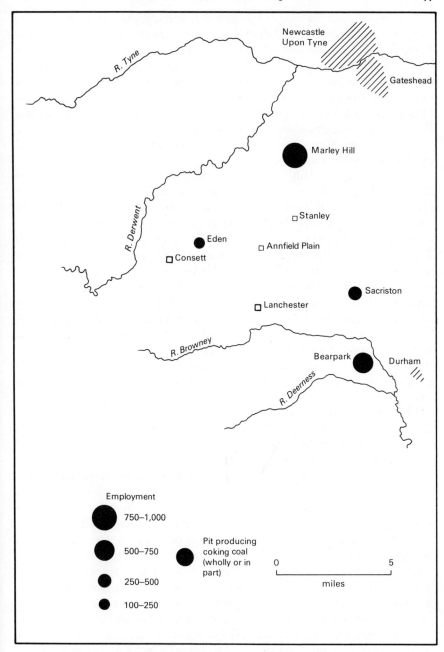

FIG. 19.1. Collieries in north-west Durham, 1977. By 1983 Eden and Marley Hill had closed; by the end of 1985 Bearpark and Sacriston had closed

million tons, though shortage of molten iron reduced effective capacity to no more than 1.5 million tons. Relatively small capital spending was believed to be capable of increasing this to 1.75 million tons. In August 1969 BSC resorted to a stop-gap measure to make fuller use of this steel capacity. It was an interesting example of changing circumstances, involving the movement of molten iron from Cargo Fleet. Up to 600 tons was transferred daily by rail torpedo car by a 63-mile roundabout route—Stockton, Ferryhill, Leamside, Pelaw, Gateshead, Beamish. Two years later iron and steelmaking ended at Cargo Fleet; by then, further improvements had brought iron and steel capacity at Consett into balance.

Not surprisingly, within the new, national context, Consett's Achilles' heel of the handling of imported ore soon came to the fore. In 1967, for an iron production of some 0.7 million tons, Consett obtained a little over 1 million tons of ore through Tyne Dock. It became clear that it would be difficult to defend this arrangement against the logic of the case for developing a major Teesside terminal. At least one brave attempt was made to turn the tables on this scheme: the Consett MP, David Watkins, suggested that BSC should expand Tyne Dock and rail ore from there to Teesside works. This idea ignored both the much bigger steel capacity on Teesside than at Consett, and the greater suitability of the Tees estuary for dredging and of its fringing low-lying lands for the layout of major ore-handling operations. In mid-1969 plans were confirmed for a terminal there to handle vessels of up to 100,000 d.w., and perhaps later of 200,000 tons. The annual throughput capacity would be 7 million tons, rising later to 10 million tons. Unloading of ore would cease at Tyne Dock.

These plans were carried through by the mid-seventies. At that time BSC signed a 10-year contract with British Rail under which the latter reopened the Washington to South Pelaw line, closed in 1970, in order to ease movement of ore from the Redcar terminal to Consett. At the works a new ore reception unit was built at a cost of £0.75 million. In this way, in the interests of lower sea freight charges and transhipment costs, a 58-mile rail haul on ore replaced one of less than half that distance. Relative to Teesside ironmaking the assembly cost situation was worsened.[3] Another indication of Consett's loss of autonomy at the same time was that it began to be supplied with considerable tonnages of semi-finished steel from Lackenby, which it further processed.

During the early 1970s BSC spent little on developments at Consett. However, at the close of the winter of 1974/5 a £12.5 million scheme was announced—a sum half as much again as that spent during the previous four years. Most of it was for the doubling of capacity for billets, a product for which Consett had long claimed very low costs by national standards. Blast furnace improvements were to receive £1 million. The Director of the

---

[3] *I and S*, Oct. 1968, p. 451; *ST* Oct. 1969, pp. 105–7; *Voice of North East Industry*, Sept. 1968, p. 9.

Teesside and Workington Group of the BSC was reported as remarking: 'You can take this as a pretty fair indication that Consett's future is assured for the next ten years.'[4] It quickly became clear that this was not necessarily the case. Even though expansion was still generally in the air, the thoughts of the central planners in the industry were moving in directions other than that of Consett. Relative neglect was to turn into purposeful contraction when worsening trading conditions began to speed the weeding-out of the less favoured works, but at no time after nationalization were Consett's more ambitious ideas congenial to central planners in the industry.

In the early years of BSC the thinking inherent in the Benson Report concerning the economic gains of focusing activities in a handful of large tidewater plants was taken up and carried further. This was a time of expansive thinking in which, in contrast with the broad schemes for reconstruction and extension under the Board of Trade Departmental Committee in 1916/17, or the development planning of the forties, the model was no longer American, but Japanese. It stemmed from an acknowledgement that since the mid-1950s the reconstructed industry in Japan had been overwhelmingly successful. So sweeping had been this achievement, so warm and unqualified was the admiration which it elicited among steelmen elsewhere, that it seems not to have been at all widely asked whether a strategy suitable for the needs of what could be regarded as a largely new steelmaking nation was appropriate in an old one. Outsiders too—including the present writer—were bewitched by the success of this model of development. It must be stressed that there can be no quarrel with the Japanese prescription for bulk steel production for a first-time, rapidly expanding, basic industrial economy. It may be appropriate too in wholly new plants elsewhere. The fault lay in the all too ready assumption of its wholesale applicability in the immensely complex structures of the world's pioneer iron and steel producer. With the benefits of hindsight some of the fallacies involved are excellently reviewed by Bryer and his colleagues.[5] The BSC development scheme was evolved in the context of an anticipated large-scale expansion of national steel capacity. In contrast with the grand strategies for British steel of 25 and 50 years before, the planners of the late sixties and early seventies had the advantage of operating in an industry whose large, integrated works were all under their own control. Earlier schemes had been moderated or frustrated by the self-interest of individual concerns; now the only way in which the zeal of wholesale reformers could be restrained was through the shortage of funds to realize all their dreams, the lobbying of interested parties through the government of the time, or the pressures of the market. All of these were to play a part.

In late summer 1971 the BSC Planning Committee projected a 1981/2 home

[4] *Newcastle Journal*, 5 Feb. 1975, 19 Mar. 1975.
[5] Warren 1969; Bryer, Brignall, and Maunder 1982: 88–99.

consumption of steel greater by 6.5 million tons, or 38 per cent, than that of 1970. It also envisaged a major increase in exports. Meeting these targets necessitated an emphasis on efficiency and low costs so that 'the planning decisions made and about to be made will bring the Corporation's costs into line with its major competitors in the course of the present decade'.[6] In June an interim report from BSC's Working Party on Major Steelworks Development had considered the possibility of at least one and possibly two large plants on greenfield sites. Nine locations throughout Britain were considered. All of them were coastal with the single exception of Corby, which was included merely as a measuring-rod for the costs of the others. From the beginning of this exercise south Teesside was considered a well-favoured possibility. BSC was already planning a steel capacity there of 11.6 million tons by 1980; a new works would involve an additional 10.25 million tons. The large expansion of North-East steelmaking did not make provision for Consett. The Planning Committee confirmed this conclusion (see Tables 19.3, 19.4, and 19.5).

By summer 1972 another study had produced a rather different list of 10 'strategic options' for 1980/1. These covered capacities for BSC ranging from 27.9 million tons of liquid steel up to a maximum of 36.4 million tons. The biggest of the scenarios involved seven integrated works, as compared with over twice that number then in operation. There was a strong in-built bias in favour of giant-sized developments and against the older works. (Bryer and his colleagues were told by an ex-senior manager of BSC that 'in the early runs of the computerised economic planning model the options *retaining* a large number of works which had not been scheduled for development were the frontrunners!' Accordingly, the planners introduced a device to penalize them. This was 'an injection of £5 a year for each ton of capacity . . . intended solely as a *convenient guide* to the total capital injection needed to keep costs (and quality standards) constant over the generality of plants' (author's emphases). Bryer, Brignall, and Maunder comment, 'We were told that making this change had the desired effect—the old plants all became candidates for closure!' (1982: 116).) In only four of the 10 strategies did Consett appear, in each instance with a steel capability of 1.5 million tons. In the other six schemes, Consett was assumed to have been closed (see Table 19.6). In any case it was reckoned that its sinter capacity would be shut down and, even more significant, that it would cease to roll heavy plate, thereby becoming wholly dependent on the production of billets. Eventually a scenario for the industry was chosen which did not include a works at Consett.

A document containing a *Ten Year Development Strategy* was presented to Parliament in December 1972 and subsequently published. By this time the national target tonnages for the early eighties had been still further increased.

---

[6] BSC Planning Committee, memorandum of Managing Director Commercial, 14 Sept. 1971.

TABLE 19.3. *1971 forecast of steel production in 1976/7 and 1981/2* ('000,000 ingot tons)

	1976/7	1981/2
Port Talbot	4.00	9.12
Llanwern	3.40	3.50
Ravenscraig	1.30	2.10
Scunthorpe	5.15	5.44
Teesside	12.65	13.30
Sheffield and other electric steel	4.30	4.43
Consett	1.20	—
Corby	1.00	—
BSC total production	33.00	37.89

Source: BSC Planning Committee, Sept. 1971.

TABLE 19.4. *BSC North-East Coast crude steel production in 1980, as envisaged in the 1971 Corporate Development Plan* ('000,000 tons)

Finished Product	Teesside	Greenfield works (probably Redcar)
Billet-based	3.67	1.61
Bloom-based	1.28	0.66
Heavy plate	2.01	1.12
Light plate	1.82	0.21
Strip mill	—	3.34
TOTAL including products not listed above)	11.60	10.25

Source: BSC 1971 Corporate Development Plan, June/July 1971.

TABLE 19.5. *Production and consumption of steel plate ($\frac{3}{8}$ in. thick or over), 1970, 1976/7 and 1981/2, as analysed in 1971* ('000 tons)

	Total demand (home and exports)	Home production
1970 (actual)	1,938	1,914
1976/7 (forecast)	2,301	2,251
1981/2 (low estimate)	2,669	2,609
1981/2 (high estimate)	2,926	2,856

Source: BSC Planning Committee, Sept. 1971.

The new plan provided for BSC to have 36 to 38 million tons of steel capacity and required an outlay estimated to be as much as £3,000 million. Consett was now classed as a marginal works, with no prospect of expansion from the current capacity of about 1.1 million tons. Not long after this changes in demand made survival even in this category look less secure. The situation in the plate trade was the first to become critical.

TABLE 19.6. *The British Steel Industry in 1980/1, as modelled in 1972*

Districts and works	Steelmaking process	Number of total of ten planning options in which plant appears	Range of plant sizes ('000,000 tons)		
			lowest	highest	average
North-East					
Consett[a]	Oxygen	4	1.5	1.5	1.5
West Hartlepool[a]/Cargo Fleet[a]		0	—	—	—
Lackenby[a]/Cleveland[a]	Oxygen, arc	10	4.45	5.95	5.72
Scotland					
Ravenscraig[a]	Oxygen	10	2.70	2.70	2.70
Hunterston	Arc	3	0.50	0.50	0.50
Lincolnshire					
Appleby/Frodingham[a]	Oxygen	10	4.70	5.20	5.11
Normanby Park[a]	Oxygen	6	1.30	1.30	1.30
Merseyside					
Shotton[a]	Oxygen	3	2.25	2.25	2.25
Northamptonshire					
Corby[a]	Oxygen, arc	10	0.10	1.40	0.62
South Wales					
Port Talbot[a]	Oxygen	10	3.10	6.10	4.30
Llanwern[a]	Oxygen	10	3.60	3.80	3.70
Cardiff[a]	Arc	3	0.50	0.50	0.50
Sheffield/Midlands & other works	Arc	10	3.90	5.70	5.15
Possible new hot metal works	Oxygen	7	3.30	6.00	4.90
TOTAL		10	27.9	36.4	33.06

[a] Works having hot metal operations in 1972.
Source: BSC June 1972.

British plate mills were under-utilized after the big extensions of the late fifties and early sixties (see Table 19.7). In 1965 the national operating rate for plate was 74.5 per cent; Hownsgill was performing at just about that rate. By 1971 United Kingdom production of plate was 73.5 per cent of the capacity planned for 1970, and over the next five years averaged no more than 66.4 per cent of this figure. Consett and indeed all north-eastern works became progressively more ill-placed to serve this low and stagnating demand as the proportion of the plate consumed in shipbuilding and repairing shrank. Those outlets which were growing in relative importance were not so strongly localized in the heavy steel districts. (Shipbuilding and ship repair took 30.2 per cent of home consumption of plate in 1954; 18.1 per cent by 1974/5.) A major centre of the plate trade which benefited from this change was Appleby-Frodingham, well located to serve Midland and South-East markets.

Apart from this marketing disadvantage, and the perennial problems associated with Consett's interior position, there were also difficulties con-

TABLE 19.7. *Deliveries of steel plate, 1971–1980*

Year	Net deliveries ('000 tons)
1971	2,831
1972	2,776
1973	3,057
1974	2,507
1975	2,121
1976	2,065
1977	2,090
1978	2,080
1979	2,282
1980	1,261

Source: Iron and Steel Statistics of Great Britain (annual).

nected with the design of the Hownsgill mill. It was capable of rolling only rather narrow plates. In the 1972 development strategy, BSC revealed that it favoured the construction of a new, wide plate mill at Redcar, thus reviving the project which Dorman Long had shelved when they entered into the co-operative rolling arrangements with Consett in 1964. By 1974 BSC were planning that this mill should have an annual capacity of 1.1 million tons, equal to about 45 per cent of the national output of plate that year. In 1975 Lord Beswick was involved in a review of the prospects both for Hownsgill, where 800 jobs were at risk, and at West Hartlepool, with twice that number. As a result both works were saved. However, the following year BSC again brought forward a scheme for a 2-million-ton Redcar plate mill. In the following year it reduced the scale, this time to 1 to 1.5 million tons. It still insisted, against the interests of the other two plants and the arguments advanced by consultants called in by the Iron and Steel Trades Confederation, that a Redcar mill was preferable, though apparently this was measured against the existing, rather than modernized, operations at Consett and Hartlepool. Even the ISTC, though against the Redcar project, favoured Hartlepool rather than Consett, referring in its report much more decisively to Hartlepool's 'enhancement', or even to the building of the new mill there. Whether the investment was at Redcar or at West Hartlepool, Consett would clearly not survive (ISTC 1977). No longer was it expanding (see Table 19.8).

At the end of the seventies acute, long-persisting depression of trade became the most serious of issues. This first threw doubt on the Corporation's expansion plans, though the planners proved extremely reluctant to revise them. Eventually, however, extension turned into wholesale contraction. The case for retention of what were regarded as marginal plants now seemed indefensible, though, as has been pointed out, it was only the Corporation's

TABLE 19.8. *Consett capacity, 1974 and 1979* ('000 tons)

	1974	1979
Coke	494	317
Iron	972	1,000
Steel ingots	1,500	1,300

Sources: Metal Bulletin, *Iron and Steel Works of the World*, 1974 and BSC 1979a.

over-commitment to massive new projects that caused it to make inadequate use of their capacity (Bryer, Brignall, and Maunder 1982: 102, 103). In 1978 BSC published a document entitled 'The Road to Viability'; it effectively marked the end of the development strategy announced in 1972.

In 1978/9 Consett operations lost £15.2 million, a level of unprofitability by no means exceptional among BSC plants. In the same year losses in Teesside works totalled £65.9 million. The difference between the Consett situation and theirs was that it was judged that, in the former instance, the problem could not be rationalized away. The plate mill suffered the first blow, from a delayed and now hurried programme of rationalization. Launchings of ships in 1978 were a little under half the level of five years before, and steel deliveries to shipbuilders and engineers were only just over 40 per cent as great. As the traditional plate outlets withered, Hownsgill's losses mounted, in the operating year 1978/9 reaching £8 million. Closure of the mill was announced on 8 September 1979. This was represented to the employees as a means of improving the overall costs of Consett operations, and, by the removal of a millstone, as a step towards improved prospects. The first point was valid, but the second, though a natural conclusion, proved not to apply. When this became clear with hindsight, there was much bitterness towards a management which was believed to have deceived the work-force.[7] The last plate was rolled on 12 October, only just over 19 years after the mill had been commissioned. With its closure there ended almost a century of illustrious history in the manufacture of steel shipbuilding materials. Some 400 jobs were lost. Even more important, the way was soon shown to be open for a direct attack on the rest of the works. Attention now focused on the critical state of the billet trade.

Demand for steel for rerolling was by this time by no means a secure refuge for a major works which had withdrawn from plate production. To a much greater extent than with the plate trade, the outlets were in midland and southern Britain. They were being reduced not only by recession, but by the inroads made by an increasing number of 'mini mills' (by 1978 there were five independent and BSC works making billets in small, unintegrated plants

---

[7] M. Upham in correspondence, May 1988.

using continuous casting). At the same time foreign competition was increasing, imports more than doubling their share of the British market in the seventies. For its billet trade Consett was already obtaining some material from Lackenby.

As the threat to Consett mounted, consideration was given to the likely consequences of drastic run-down or outright closure, and also to the possibilities of alternative employment. Some of the impact studies were little more than impressionistic; others were statistically based, though still often speculative. As early as 1968, the local MP, David Watkins, had reckoned closure of Consett would mean the loss of £25 million annual spending power in the North-East region. He had sketched what he saw as a catastrophic situation, at least sub-regionally: 'every business in North West Durham would face bankruptcy. The waste of social capital, in houses, schools and other facilities, would be incalculable. The human suffering could not be measured.' Yet in the same year an attempt was made to measure these regional ramifications.[8] J. R. Atkinson's first impact study on behalf of Durham County Council had been made in January 1967. It was updated some 18 months later. Atkinson reckoned that the multiplier effect of closure would involve a total loss of 11,600 jobs in north-west Durham, with 920 jobs, from direct Consett employment, lost on Tyneside as well. Two years later another analysis put the probable loss in north-west Durham at 14,000, with additional effects not only on Tyneside, but extending as far as the Tees (Roberts 1970). In fact, instead of the disastrous, sudden closure then anticipated, there was a progressive erosion of jobs throughout the seventies, this, too, with its snowball regional impact. During the same period new industries were introduced, both to replace at least some of the male jobs lost in steel and coal, and to provide new possibilities for female employment. In the fifties, Consett works had been the only manufacturing plant of any size in the area which in the reorganization of 1974 became Derwentside; by 1978 it employed no more than 34 per cent of the manufacturing work-force there.

During 1979, the consultants Coopers and Lybrand and the Building Design Partnership were commissioned by BSC Industry Ltd., the European Commission, Durham County Council, and Derwentside District Council to study the area's situation and prospects. What was described as the 'urgent need to mount an effective regeneration strategy' should Consett works close caused the analysts to complete the substantive work within three months, and in November *Derwentside: A Strategy for Industrial Regeneration* was published. The job losses already suffered by the run-down of the works were analysed. The consultants then went on to recognize that they had to take into account a 'hypothetical total closure of the steelworks' with a further loss of 3,300 jobs, 'but not before the mid-1980s' (Coopers and Lybrand 1979: 16).

---

[8] Watkins 1968; Atkinson 1968.

However accurate the econometric model with which the consultants worked, their timetable proved wildly inaccurate.

Until now, as the consultants' words implied, the period over which any final run-down of Consett might occur had been reckoned as five years. This was regarded as a period over which it might reasonably be expected that the necessary adjustments to the town and regional economy could be achieved. Now time began to run out much more quickly. In the month following the publication of the Derwentside study, and so near Christmas that publicity and discussion were muted, the BSC revealed sweeping rationalization plans for the whole of the industry. These were contained in a document entitled *The Return to Financial Viability: A Business Proposal for 1980/81*. This stressed the necessity for major reductions in national steel capacity in order to respond to a new market situation which had been created both by a decline in the consuming industries and by increased imports encouraged by high rates of inflation and by a high sterling exchange rate. The Corporation now set itself a capacity target of no more than 15 million tons—less than half the figure which under three years before it had still held on to as the figure for the mid-1980s. Nationally it was expected that 50,000 men would be made redundant.

In the course of 1980 the new plans brought a response from the Iron and Steel Trades Confederation in the form of a *New Deal for Steel*. Not unnaturally, but also persuasively, ISTC maintained that the BSC document had been a hasty response to a new government's reduction of its cash limits, and to the requirement that the industry become self-financing as quickly as possible. Such restrictions, ISTC argued, meant that adequate investment could not take place. In its judgement the so-called BSC strategy was in fact only a tactical development, for it contained no long-term plans, and ignored a range of possibilities which could save a number of the threatened works.

Meanwhile, whatever the rights or wrongs of the matters in dispute, the economic situation of the BSC, and thus its attitude to the need for and the urgency of rationalization, was not improved by the thirteen-week national strike during the winter of 1979/80. This worsened the outlook for works already regarded as marginal operations. It was BSC's *The Return to Financial Viability: A Business Proposal for 1980/81* which directly anticipated the early closure of Consett. At last, within the context of a national industry, investment on Teesside was to bring about the elimination of a works which had successfully resisted competition from Teesside for 130 years.

20

The Progress to Closure, 1980

FOR almost six months after Consett's closure was mooted in December 1979, there seems to have been no consultation on Consett's future between BSC and the employees' representatives or other interested parties at either the national or the local level. Certainly there is no public record of such discussions. Threatened so often, it began to seem that Consett might after all escape again. In spring 1980 the town was not without signs of optimism and even of development. The pedestrianized shopping precinct was reasonably busy, though there was already evidence of high teenage unemployment in the numbers of teenagers present in the area on a workday afternoon. The flower-beds in the central areas were gay with colour. In the nearby large central square, which had for so long acted as an open-air bus station and car-park, construction was underway to provide the town with a covered coach station. The community had clearly not given up hope. Even so, there loomed in the background the threat of disaster, and it was all too clear that the struggle to survive might be harder than at any earlier time.

Managerial opposition to closure of Consett was shortlived and ineffectual. A week before the publication of *The Return to Financial Viability*, the Consett General Manager was transferred to Teesside. This removed a potentially powerful leader of local opposition to those proposals—though soon there was much resentment resulting from a belief that he had deceived the men about the longer-term prospects when they had been persuaded to tamely accept the plate mill closure. There was only a brief effort to mount a defence on the part of the pre-nationalization management of Consett Iron. In a letter to *The Times* Roland Cookson, former Managing Director, advocated a policy of holding on, as Consett managements had done in the past. If this policy was followed, then, as before, when trade conditions improved Consett would again take its place as a valued contributor to the national steel supply. Cookson argued that this position was still a viable one, and that there was always a place for 'the smaller, medium-sized producer in any industry'. It was fallacious to suppose 'that all requirements could be met from one or two giant plants'.[1] In short, he maintained that Consett could still play the part that the planners of the BSC strategy had foreseen for it eight years earlier (see Table

---

[1] R. Cookson to *The Times*, 29 Feb. 1980.

20.1). The rearguard action fought by the Consett steel unions was far more protracted and important than these managerial protests.

In summer 1980, encouraged by successes in South Wales in securing acceptance of very large reductions in labour forces by means of generous redundancy payments, BSC decided to act on Consett. Early in June the Personnel Director of the Corporation's Teesside Division wrote to the Divisional Officer of the Iron and Steel Trades Confederation to announce the timetable for closure. His letter, the most momentous in the history of Consett, read:

CONSETT WORKS—TEESSIDE DIVISION

I am writing to advise you of the Corporation's proposals concerning future operations at the Consett Works. The information below has also been notified to the National Officials in accordance with normal procedures.

The intention to close Consett was included in the 15 million tonne strategy statement announced by the Corporation's Board in December 1979. This decision was taken against the background of falling demand for Billet and Billet-derived products which have been particularly hard hit by the decline in the motor vehicle and engineering sections of the UK industry. Despite the high level of export achieved in the latter half of the 70's the estimated excess capacity over demand is forecast to increase during the first half of the 80's. Consequently under recovery of fixed costs already being incurred will be further aggravated unless urgent action is taken to reduce the Corporation's Billet capacity.

After due consideration of alternative options the Corporation has concluded that the most economical approach is to close the Consett Works and re-allocate the declining load to its recently modernised Billet producing mills in the Yorkshire and Humberside Division. The Corporation is confident that the transfer of the load can be effected without incurring any commercial disadvantage as the alternative Billet manufacturing facilities can absorb the totality of the Consett order load without further capital expenditure.

On Thursday 12th June the local workforce will be presented with copies of the closure brochure for the Consett Works. The document (copy enclosed for information) confirms the Corporation's intention to cease operations at the works by the end of September 1980.

I would be grateful if you would regard the above information as confidential until that time.

Yours faithfully,

M. D. WARD
Director—Personnel
Teesside Division

The letter was posted from Steel House, Redcar on 11 June. Its destination was the ISTC Divisional Office on Marton Road, Middlesbrough, not more than 6 miles away. It was not received there until 13 June. On the previous day, the Managing Director of the Teesside Division had informed union leaders gathered in the Civic Hall, Consett, that the works would close at the

TABLE 20.1. *Pig iron and steelmaking at Consett and Teesside works, 1953/4–1981 ('000 tons)*

	1953/4 (average production)		1965 (production)		1973 (capacity)		1981 (capacity planned in 1972)[a]
	Pig iron	Steel	Pig iron	Steel	Pig iron	Steel	Steel
Consett	572	652	730	1,045	972	1,500	1,500
South Durham and Cargo Fleet	477	750	1,004	1,430	1,321	965	nil
Dorman Long	1,242	1,811	1,533	1,915	2,642	5,250	5,950[b]

[a] Highest of 10 possible figures.
[b] Plus a possible 6-million-ton new plant.

Sources: Iron and Steel Board, *Annual Reports and Development in the Iron and Steel Industry: Special Report* 1957; BISF 1966; and BSC Planning Committee papers, 1972.

end of September. As he spoke, the hall was picketed by men campaigning to save the plant. By this time, though the location was recognized as far from ideal, the gravamen of the case against Consett centred on its scale and equipment, which in BSC judgement rendered it incapable of surviving in an extremely difficult market situation. Britain had a number of sub-optimal plants from the point of view both of size and location. They were competing with each other for survival. BSC was determined to seize the opportunity to sweep most of them out of existence.

In spite of considerable capital spending in both the last years of independence, and to a lesser extent since nationalization, much of the plant and equipment at Consett was now inferior to that elsewhere in the BSC. At the end of 1976 the 52-oven coking plant at Consett was, with the single exception of the installation at Brookhouse, near Sheffield, the smallest of the 14 carbonization units in the Corporation. These ovens were, it is true, modern, dating only from 1969, but there is some evidence that their design was rather conservative. The two sinter plants had been installed in 1953 and 1954, and had not been modernized. The three Consett blast furnaces were built in 1941, 1947, and 1950 (though remodelled in 1977, 1974, and 1976 respectively). They were of 6 to 7 metres hearth diameter; the BSC had 14 smaller furnaces, but 24 which were bigger. In its total ironmaking capacity Consett was in 1978/9 sixth in size of 11 British plants; all except Ravenscraig of the bigger works were at least double the size of Consett. At that time the average capacity of the oxygen steel converters in Britain was 193 tons; the two Consett vessels were now rated at 132 tons, though 140 tons per heat was being obtained from them by 1979. The blooming, slabbing, and billet mills had been installed in 1953 and had not been extensively altered since then. Again, however, Consett had done well with inferior plant, the hourly output of the billet mill being higher than that of the unit at South Teesside, though the latter was 11 years younger. Scale deficiencies did not apply to the finishing operations. By contrast, iron, steel, and primary rolling mill operations at Consett were quite clearly too small, for it was in these departments especially that the best practice of the metallurgical world had moved furthest in size. Although Consett recorded low ironmaking costs in the late seventies as compared with most British plants (see Table 20.2), these figures were to some extent unrealistic. The very low sinter cost is a reflection of the old, written-down nature of the units, and the low depreciation charge per ton of iron points to the age of the furnaces. Extended life for the Consett works would have required spending on the capital account for both these sections, and so eventual removal of any Consett advantage in ironmaking costs. It will be noted that, with the exception of Corby and Shotton, Consett had the highest ironmaking costs of the 10 plants listed apart from the costs of depreciation and sinter.

By the late 1960s it was clear beyond dispute that for ordinary grades of steel, if made in a conventional blast furnace–converter–primary mill–

TABLE 20.2. *Molten iron costs at Consett and other BSC plants, 1978/9* (per ton)

Blast furnace plant	Total cost	(a) Sinter	(b) Depreciation	Costs per ton apart from (a) and (b)
	£	£	£	£
Consett	69.47	16.98	1.09	51.40
Cleveland	67.43	26.76	3.22	37.45
Redbourn	74.17	28.35	2.42	43.40
Normanby Park	74.71	28.94	2.39	43.38
Appleby	75.17	28.35	2.63	44.19
Port Talbot	76.42	25.02	2.33	49.07
Llanwern No. 3	76.90	25.74	2.65	48.51
Shotton	78.93	20.71	1.04	57.18
Ravenscraig No. 3	79.57	36.49	3.52	39.56
Corby	94.98	26.60	2.09	66.29

Source: BSC Standard Cost Statistics, 1978/9 (unpublished).

finishing mill sequence, a new, higher level of capacity had become desirable, and that for such extended plants the attractions of tidewater location were greater than ever. Japan already had some blast furnaces with a 10,000 ton per day iron capacity. For security of iron supply a 2-furnace iron plant is desirable (though as events were to prove it was not impossible to manage with one furnace). Two 10,000 ton per day furnaces will give an iron capacity of about 7 million tons a year. This, along with the integrated work's own arisings of scrap, would support a steelworks of some 8 million tons, perhaps in the form of a 4-vessel oxygen converter shop, a high-capacity primary rolling mill, and a continuous casting unit as well. This sort of plant is the product of the operational logic of the new technologies. It is true that perfectly satisfactory economies of scale can be secured at levels well below these, but whether they could be won at anywhere near the size of Consett works is at least highly doubtful. Slightly smaller 'giant' furnaces, say two of 8,000 tons per day, or one of 10,000 tons combined with two existing smaller furnaces, would still push the level required well beyond any which had been seriously contemplated for Consett. Although it would have increased the overheads, a thoroughly reconstructed existing furnace plant might have given more favourable costs of operation. Another possibility might have been to scrap the furnaces and the LD converter plant and to install electric furnaces dependent on scrap. This might have been a rational solution from an economic point of view, and certainly socially much to be preferred to complete closure of iron and steel operations at Consett, but there were good reasons why it seems to have received no attention. BSC was adopting crisis measures to deal with a critical immediate situation. The urgency was to cut costs, and the easiest way to do this was to close plants: longer-term, positive planning could not be contemplated. Moreover, there were other problems which in these circumstances ruled out Consett as a continuing operation. Any deficiencies in

TABLE 20.3. *BSC billet capacity, 1980* ('000 tons)

	Billet capacity	Billet grade (c = commercial, s = special or alloy)
Consett	750	c
Appleby (Scunthorpe)	2,100	c
Normanby Park (Scunthorpe)	820	c
Aldwarke (Rotherham)	900	s
Stocksbridge/Tinsley (Sheffield)	875	s
Templeborough (Sheffield)	760	s
Bilston (Staffs)	130	s^a
TOTAL	6,335	

[a] Dependent on ingots from Sheffield.
Source: BSC 1980a.

equipment there could be corrected; on the other hand, from the perspective of central planners, especially when operating in beleaguered markets, its locational unsuitability ruled it out as a works which should be saved.

In June 1980 BSC presented its case against Consett in a 24-page document *The Case for Closure*. Now that the plate mill had been abandoned for eight months, *The Case for Closure*'s most important point concerned the collapse of the markets for billets, and Consett's inability to justify a place among the continuing suppliers of that reduced demand. In 1970/1 United Kingdom consumption of billet products had peaked at almost 10 million tons, of which BSC had supplied half. By 1979/80 demand was down to 8 million tons and still falling; the BSC share was only 35 per cent. In short, home outlets for billets from Corporation plants fell from about 5 million tons to 2.8 million. As suggested above, this declining share of a shrinking market was the outcome of three conditions: extended production by some consumers of their own billet requirements, as in the case of GKN; the emergence of new private steel firms, such as Manchester Steel or Sheerness Steel; and increased imports. Meanwhile BSC seemed to be pricing itself out of the business. (Between 1975 and 1980 prices for various types of billet increased by between 95 and 115 per cent, whereas the BSC costs for 'commercial' (that is, non-alloy or non-special grades) went up by 165 per cent.) By 1980, although it had already closed billet and associated steel capacity at its Hallside and Cleveland works, the Corporation still had a billet capacity of 6.3 million tons, 3.5 million tons in excess of the current level of demand. This level it now expected to fall even lower (see Tables 20.3 and 20.4). British Steel set itself the target of reducing billet capacity by about one-third to meet the demand, already largely concentrated in the south, that it anticipated for the mid-1980s (see Table 20.5). Of the surviving producers in the business, Consett was one of the smaller ones; in the 'commercial' grades, the smallest.

BSC argued that closure of Consett would enable it to bring billet capacity a

TABLE 20.4. *BSC billet and steel capacity and demand, 1979/80 and 1985* ('000,000 tons)

	1979/80		1985 (anticipated)
	Capacity	Demand	Demand
Billets	6.3	4.9	4.1
Liquid steel equivalent	8.0	6.2	5.0

Source: BSC 1980a.

TABLE 20.5. *The regional distribution of British demand for steel plate and billets, 1965 and 1980* ('000 tons)

Region	Plate deliveries for home use		Billet-based consumption
	1965	1980 (forecast)	1980 (forecast)
North	503	760	360
Scotland	343	350	440
Yorkshire and Humberside	112	100	920
North-West	170	200	1,230
East and West Midlands	446	500	1,450
Wales	53	100	520
Northern Ireland	44	60	60
South-West	28	30	190
East Anglia	50	30	130
London and South-East	153	170	1,220
TOTAL	1,902	2,300	6,520
Total, north of Wash–Merseyside line	1,172	1,470	3,010
Total, south of Wash–Merseyside line	730	830	3,510

Notes: The 1965 figures represent a regional breakdown of the BISF consumer census for the UK. This amounted to 74 per cent of the national total. 1980 figures are BSC Planning committee projections made in 1971.

good way towards balance with demand (see Figs. 20.1 and 20.2). Already the expansion of the Redcar and Ravenscraig works had removed much of the need for Consett production for inter-works transfers of semi-finished steel within the Corporation. It was now proposed that Consett billet orders should be reallocated: orders totalling 3,000 tons weekly would go to Normanby Park, and 10,000 tons to Appleby-Frodingham. These plants would even supply some of the billets required by the rolling mills at Jarrow and at Lackenby, which had hitherto come from Consett. Lackenby works would take over from Consett the small business in slabs for the plate rerolling operations at Redheugh on the Tyne. All in all BSC estimated that closure of Consett and

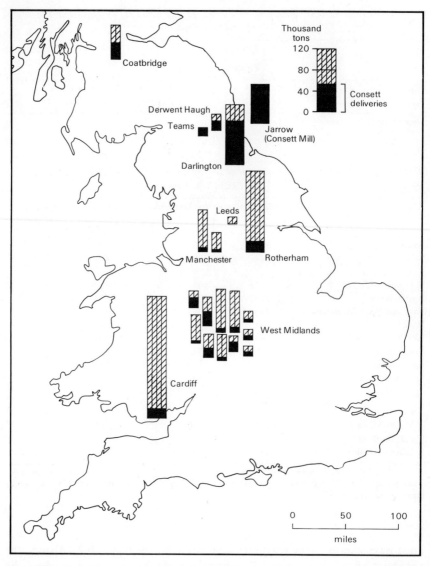

FIG. 20.1. Consett and total deliveries of billets and billet slabs to various destinations, 1978/1979

the introduction of these alternative arrangements would bring annual savings of £40.5 million. Looked at from the Consett perspective, it seemed that a works which, against the better judgement of its own management, had less than 30 years earlier been cajoled into a large role in the billet trade was now to be sacrificed to bring home demand and capacity for billets into closer

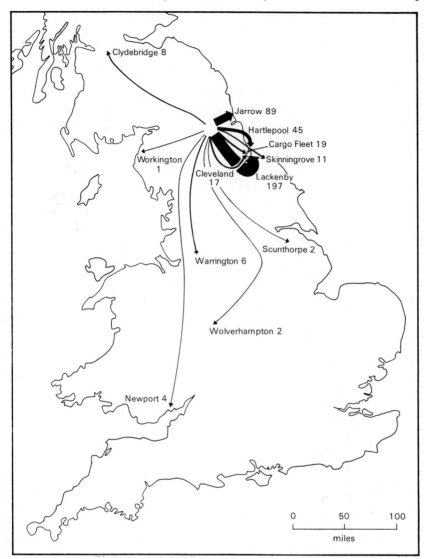

FIG. 20.2. Consett deliveries of slabs, blooms, and billets to other BSC works, 1979 (thousand tons). Out of total deliveries of 680,000 tons, deliveries to BSC works totalled 401,000, of which only 23,000 tons were delivered outside the North-East

relationship. In the course of the spring and summer of 1980, markets for billets declined even faster than had been anticipated; this strengthened BSC's resolve to close Consett (BSC 1980*b*).

In July the trades unions at Consett published a refutation of the BSC case in a document which they boldly entitled *No Case for Closure*. It was in many ways a more impressive document than the BSC one, though subsequent evidence threw doubt on some of its statements. The billet situation was analysed, and then attack moved on to other aspects of Corporation development policy. BSC, it pointed out, sent 400,000 tons of Consett products to its other works, and 280,000 tons went direct to customers. In the absence of evidence the unions questioned whether Consett was in fact a high-cost producer. They argued that Consett billets were very highly regarded by their customers, some of whom were said to maintain that to make billets of equal quality BSC would have to invest heavily in their other works, which would push up production costs. Much was made of Consett's high productivity. It was said that increases in electricity charges would mean that customers who had been turning out more of their needs for steel in electric furnaces or who by means of small continuous casting units, had supplied their own billets, might be forced to return to purchasing both from bigger, integrated works.

In the second part of *No Case for Closure*, entitled 'The Alternative to Closure', these points were developed further. At the same time there was criticism of the BSC plan for the North-East where, should Consett close, all operations would be dependent on the one giant blast furnace at Redcar, and the single bulk steel plant of Redcar/Lackenby. (Not only the operational, but also the economic rationale for this furnace has been questioned. At the Clay Lane furnaces, Teesside, which were closed to make way for Redcar furnace, the 1978/9 standard costs were £67 per ton of iron. The *projected ideal* cost at Redcar was only marginally less at £65 per ton—a saving won by a capital outlay of £400 million (Bryer, Brignall, and Maunder 1982: 184).) *No Case for Closure* championed Consett as a 'back up' works for BSC Teesside, as a continuing source of supply for the rolling mills at Skinningrove and Jarrow, and as a leader in an aggressive counter-attack on the import trade in billets, which overall they believed might recapture 150,000 tons of the market. It was suggested that Consett could sell 900,000 tons of semi-finished steel, or much more than it had recently been providing. This was justification for the retention of a plant capable of 1 million tons of ingot steel a year (see Table 20.6). A further stage might involve the improvement of the Consett billet mill in order to aim for the market which lay between the bulk, 'commercial' outlets and those for higher-grade material. Consett's experience with the use of oxygen furnaces was said to be relevant in this respect, though neglected by the BSC. It was intimated that Consett was the only works in the country which might move 'up market' in this fashion.

The ISTC document quoted the Corporation's figures to support a claim that Consett production costs were the lowest in the country. (In fairness to BSC it must be stressed that these figures ignored the fact that the burden of capital expenditure at Consett had already been largely written off, whereas in those works where large sums had been spent in recent years cost figures were heavily weighed down by this investment.) The progress which had been made

TABLE 20.6. *The ISTC strategy for Consett semi-finished steel, July 1980*

Proposed purchaser of semi-finished steel	Proposed deliveries ('000 tons)
UK market	500
BSC light section mills	140
Teesside, as back-up supply	250

at Consett in improving productivity was undoubtedly impressive. In August 1979 the General Manager had called for new, higher levels of efficiency of operation, and these had been attained in the next few months, in the blast furnaces, the primary mill, and in billet production. As the plate mill closed, losses at Consett turned into increasing profits: £24,000 in September 1979, £110,000 in October, and in the following month £300,000. BSC liquid steel output per man year was in 1979 a mere 140 tonnes, as compared with 238 tonnes in West Germany. The Consett figure was 240 tonnes, though this achievement must be placed in perspective, for, as a producer of semi-finished steel only, the labour force was now small in relation to the tonnage it obtained and sold. The ISTC case was summed up in the claim that now, as in the past, Consett was renowned for experience, reliability, and flexibility.

The BSC responded on 6 August, when it published an appraisal of what it called the ISTC 'brochure'. The new BSC document was considerably fuller and more authoritative than the Corporation's original one, *The Case for Closure*, which can only imply that originally it had been assumed that a sketchily outlined argument would suffice. The Corporation now pointed out that some other British works had achieved higher levels of productivity than those recently reached at Consett—at Normanby Park, for instance, the level was 26 per cent greater. The union criticisms of relying so heavily on Redcar/Lackenby and particularly on the single Redcar furnace were rejected. (On this point, however, events in the mid-1980s were to go a considerable way towards vindicating the union case.) The BSC response on the questions concerning the role of Consett in the billet trade seems to the outsider (inevitably dependent on evidence provided from within the industry) more effective. BSC confirmed its earlier gloomy prognostications of the demand trends for billets, and denied that Consett billets had any special qualities which rendered them less liable to rejection than those from rival makers. The Corporation went further and proved, at least from the purchases made by GKN in 1979, that Consett was, as regards quality of product, a less successful billet maker than Normanby Park, and much less so than Appleby-Frodingham. (Even so, Normanby Park was closed by BSC in 1981.) As a result, the objectivity of the ISTC survey of billet customers which suggested a strong preference for Consett products was questioned, BSC pointing out that its own reports led to no such conclusion. Finally, denying that Consett was especially well fitted to move up-market with billets made from oxygen converter steel, the Corporation indicated that other works were as adequately skilled and better equipped to explore this field of possible advance. The 17-page BSC document concluded:

Having given detailed consideration to the Trades Union brochure *No Case for Closure* the Corporation cannot find adequate substantiation for the conclusions and recommendations it presents or justification for departing from the decision to effect a permanent reduction in billet product capacity by ceasing operations at the Consett works from the 30th September 1980.

A week after the publication of the second BSC document, on 12 August, a three-hour meeting between the ISTC and BSC's regional senior management failed to weaken the latter's resolve. An attempt at intervention by the Steel Committee of the TUC proved equally ineffective. Meanwhile publicity on behalf of the rescue campaign continued. Marches were held both in Consett and at Westminster, and a petition was sent to Whitchall. The results were insubstantial. While this was going on the force of the opposition to closure was weakened by the unexpected readiness of a large number of the Consett workers to accept the redundancy payments which BSC was offering. There was a statutory figure plus a 50 per cent supplement, but even so there seem to be irreconcilable differences in the various figures quoted. One account suggests that the payments made totalled £12.4 million, which would work out at an average of £3,350 per man, but another indicates an average, including an EEC contribution, of £7,000. The whole system was bitterly denounced by the General Secretary of the ISTC as 'fool's gold', though the willingness of the men to make the best arrangement for themselves and their dependents was understandable. Whether his judgement was fair or not, the inducement had the desired result. In the event Consett was to close three weeks earlier than BSC had scheduled.

Consett's steelmaking career ended on Thursday, 12 September 1980. From various parts of the works workmen converged on the basic oxygen converter LD3 in order to be present when the last metal was made. The vessel carried two painted slogans—'Teesside cuts off its right arm' and 'RIP 1864–1980'— but as the molten iron was charged and converted, the men stood quietly. When the steel was poured, not only were samples taken as usual to the laboratory, but also small moulds were filled so that each man could take home a memento. A piper played a lament through the works. In this emotion-charged atmosphere, 140 years of commercial activity came to an end (see Fig. 20.3).[2]

There was a brief coda to the long history of Consett in steel. A few days before the closure a consortium entitled the Northern Industrial Group wrote to BSC and communicated to the press concerning its interest in buying the

[2] Elliot 1982; Price and Wade 1983.

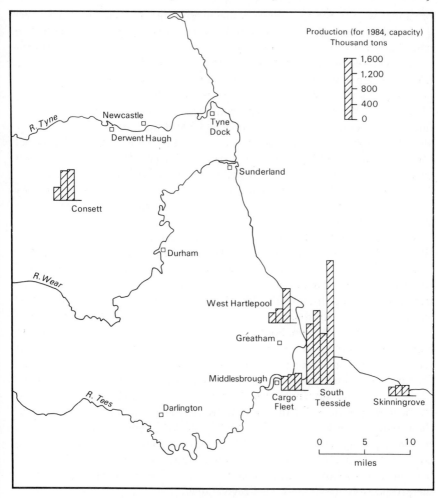

FIG. 20.3. Steel production of North-East Coast works, 1937, 1945, 1967, and capacity in 1984. Production at West Hartlepool in 1967 includes both the old 'North' works and the new 'South' works at Greatham

plant. It soon became known that for a few weeks the group's representatives had been receiving encouragement from both the Department of Industry and the Prime Minister. A week after the closure a little more was revealed about the consortium. Eleven northern companies were involved; they were both public and private, and engaged mainly in the stockholding or fabrication of steel products or in supplying the steel industry. It was said that they had a combined annual turnover of £700 million, though that impressive figure was seen to be not particularly relevant when it was also revealed that the amount of capital with which they could immediately support their bid was a mere £1.5 million. Beyond that they were known to the public only through their consultant. However, some indication was provided of their aspirations for the works. They envisaged reopening it within two months of an agreement with BCS. The aim in the first instance was to produce about 600,000 tons of billets annually, both for the home market and for the Continent, particularly Holland and West Germany. Provisional customers were said to be already lined up. As compared to a labour force of 3,700 at the time of closure, the new work-force was to total 2,700, so increasing labour productivity to 320 tons per man year, almost 130 per cent more than the current British average. It was suggested that there might be greater union participation in the operations. More use would be made of Tyne port facilities, there were promises to buy only British coal, and there were even ideas concerning reactivation of the Hownsgill plate mill.

On 18 September the members of the Northern Industrial Group met the BSC for the first time. The Corporation agreed that during the negotiations they would keep heat on in both the blast furnaces and the coke ovens in order to avoid damage to them. At the same time, however, BSC set deadlines for the termination of the discussions. Their first date was Sunday, 21 September, and they then agreed to a week's extension at an additional maintenance cost which was put at £0.25 million. BSC went on to require that the consortium reveal the identity of its members and establish their financial credentials. It asked the group to agree to reimburse the extra week's maintenance costs. Finally, it required repayment of the severance and redundancy payments which it had made for those of its former workers who would be re-employed by the new operators. Whether or not this was the intention, the imposition of these extra financial burdens effectively killed off the hopes of what might have become a cost-effective rival for BSC in a major section of its business. The last-minute rescue bid, whose serious intent many questioned, fizzled out. The blast furnaces and the coke ovens were allowed to become cold, and with that the prospect of any Consett steel revival died.

21

The Jarrow of the Eighties?

IN its document *The Case for Closure*, BSC recognized that the ending of steelmaking at Consett Iron would have a serious regional impact. At that time 96 per cent of the work-force lived in Derwentside, where unemployment was already about twice the national average. A few weeks later, on 9 July, when 650 Consett workers marched to Westminster to urge on Parliament the case against closure, the demonstration itself was eloquent of the sad effects of industrial rationalization. The procession, led by the Consett Junior Brass Band, included a mixture of schoolboys, apprentices, and school-leavers who had been unable to find jobs. A petition was handed in at 10 Downing Street by two young people. The district MP, David Watkins, speaking that day was yet again dramatic in his imagery. He warned that if the closure went ahead Consett would become 'the Jarrow of the '80s'. By the end of September, with the closure completed, and the bid of the consortium having failed, there seemed every prospect that his prophecy would be realized in the collapse of the fabric of the local economy and society.

Long before the closure of the steel works, employment in north-west Durham had been shrinking, while the situation in County Durham as a whole had been stabilized (see Table 21.1). Consett Iron had been a centre of relative job security while the numbers employed in the district's collieries had tumbled. During the summer of 1980 the last pit closed; the 150 men who lost their jobs there were the survivors of a mining work-force which 30 years before had totalled 17,000. Even the new industries which had been introduced to help replace coal-mining were now rationalizing their own operations and thereby laying off workers. In autumn 1980 Ever Ready sacked 134 of its workers at Stanley, though they more than compensated for this by revealing that they would move their Headquarters and Research and Development Division to Tanfield Lea with 300 new jobs. The closure of the major Annfield Plain bearings plant of Ransome, Hoffman, Pollard with the loss of employment for 1,250 was announced in November. Altogether in the course of 1980 and 1981 there were 6,754 job losses in Derwentside district. Of this total the redundancies at Consett steelworks made up 55 per cent and the combined Ransome, Hoffman, Pollard and Ever Ready losses, 20.5 per cent. In April 1979 the unemployment rate in Consett/north-west Durham was 10.6 per cent. By September 1980, when the works closed, the rate was approximately 15.5 per cent, compared with a national figure of a little over 8 per cent.

TABLE 21.1. *Employment in County Durham and in Derwentside district, 1951–1979 and 1986 ('000s)*

	1951	1961	1971	1975	1979	1986 (spring estimate)
County Durham	238.8	237.2	227.4	234.5	235.0	232.7
Derwentside district	39.3	39.5	30.1	29.8	29.0	31.2

TABLE 21.2. *Population of the Derwentside district, 1951, 1961, 1971, 1976–1983, and 1986*

Year	Population ('000s)
1951	102.4
1961	99.5
1971	92.4
1976	91.4
1977	90.6
1978	90.0
1979	89.7
1980	89.5
1981	88.8
1982	87.9
1983	87.7
1986	86.7

Source: Durham County Council, *County Durham Structure Plan* (annual report).

In July 1982 the national rate exceeded 13 per cent, but that of Derwentside stood at 28 per cent. In addition to these unfavourable indicators, the population of the area had been declining for many years (see Table 21.2).

Of the 3,700 who lost their jobs when the steelworks closed, 80 per cent lived in Consett itself. Even so, as long anticipated, the ramifications of the closure had wide effects on the remainder of Derwentside, and indeed on much of the rest of the region. The works had provided 16.7 per cent of Derwentside district's rateable value; the rate income was lost at the same time as the calls upon it were increased. The multiplier effects of the closure were hard to identify and still more difficult to quantify, but in summer 1980 some of the more pessimistic commentators anticipated that a direct consequence would be the loss of an additional 2,000 jobs in railway employment, in local haulage, and in shops and services. Such an estimate may have been made in an objective enough fashion, and may have been correct, but it does seem to be on the high side. It was now reckoned that up to 1,000 jobs in Durham coking-coal collieries, mainly in the east of the county, and in limestone quarrying, fluorspar mines, and in Tyneside engineering would disappear as a direct result of the closure. The financial balance sheet of the effects was confused, involving a mixture of short- and longer-term conditions. Redundancy payments were probably £12.4 million. Unemployment benefits, rent and rate rebates, and reduced tax and National Insurance contributions meant a further net saving to the area of £14.6 million, though not all of this came at once, and much of it was quite theoretical, as there were no longer any earnings from which the National Insurance and the income tax could be derived. To government this was a net cost. The writing-off of capital at the works was valued at £30 million and the necessary outlay to bring in sufficient new industry to keep unemployment down to the level of winter/spring 1980 was estimated at £80 million. From the national perspective the total cost of the closure of Consett therefore amounted to about £137 million. A portion of that would be in the form of recurrent costs. In the light of such figures the insistence that BSC should balance its books may seem to have been a remarkable parochialism. On the other hand it is arguable that without such a policy the whole of the Corporation would have been penalized in international competition.

The redundancy payments bolstered the local economy for a time. Much of this money went into the financing of a rash of small businesses, for some of which, not surprisingly, the long-term survival prospects seemed doubtful. However, by Christmas 1980 there was an air of bustle and even a verisimilitude of prosperity about the town, as those who had not turned their redundancy payments into capital investment spent them. Even so the hurt went deep, and it now had become clear that it was not only material but also psychological. There was first of all a sense of loss of the very structures of everyday experience, not only of the regular routine of daily work, but also of the physical context of life in a town which had in its landscape as much as in its economy been dominated so long and so completely by Consett Iron. Such a shock is only possible in a long-established, one-industry, isolated community. One old employee summed it up very well:

I've thought about it and thought about it and I just can't imagine it. I'm sixty. All my life the works have loomed over this town. My dad worked there; his dad before him. As a kid I went to sleep with the sound of those mills running. I just can't imagine this town without a steel works.[1]

In the months after the closure other problems began to come to the surface. Aimlessness and frustration became common features of Consett's collective identity. In December 1980 the local clergy felt so concerned as to write an open letter to the Prime Minister about the state of society in the town. They

[1] *Financial Times*, 19 June 1980.

observed that there was an 'almost touching faith' that something would turn up, but already, only three months after the closure, there was evidence of the 'human degradation' of once proud workmen, and of 'increasing tensions and problems' in many aspects of personal and community life. Men had travelled as far as Birmingham and Bristol to look for jobs, usually unsuccessfully. There were house-owning families which, if they contemplated moving to work in other regions, could not sell their houses except at ruinous loss in relation to house prices elsewhere, and council house inhabitants who could not move because no one would exchange with them. Emphasizing the real desire for work among men who were emphatically not wishing to live on the charity of society, the Consett clergy continued:

> We ask you publicly to express a real understanding of the plight of men and women who have no jobs, who see little chance of getting jobs anywhere, and to whom the whole purposelessness of weeks, months, years perhaps, of unemployment will sap their spirits to the point where all hope is lost, all self-respect extinguished.[2]

In the light of these problems, the dangers of a Jarrow-type situation seemed clear and insistent. It would be quite wrong to underestimate the seriousness of the difficulties which Consett society faced, and particularly the affront to the dignity and the morale of labour. Even so the circumstances were very different from those which had blighted not only Jarrow but so many other industrial towns in the inter-war years. A range of sources of hope and help were available which those affected by earlier crises could not call upon. In the first instance, by virtue of its prominence in the urban hierarchy of west Durham, Consett had been chosen by the County Council as one of the 12 'major centres' in the new county structure plan. Stanley, nearly 6 miles away to the east, was the only other settlement designated a 'major centre' in the north-west Durham sub-region (Durham County Council 1981). In short, though it had lost its staple industry, Consett was to be provided with the industrial estates, the public services, the retailing facilities, and the infrastructure to make it a focal point in the replanning of the county. In 1985, with a resident population of 30.6 thousand, Consett/Castleside/Leadgate was the fourth biggest urban focus in the county after Darlington (85.5 thousand), Durham city (39.6 thousand), and Peterlee/Horden (30.8 thousand). Cooper Lybrands and the Building Design Partnership had already recommended industrial promotion and the provision of more industrial sites, manpower and social developments, and the improvement of communications and the environment. All this would entail large-scale expenditure, a total estimated by the consultants at £75 million over a five-year period. Fortunately, a number of sources of finance soon began to provide the necessary assistance.

The steel nationalization act of 1967 required BSC to pay attention to the social consequences of its actions. In response to this, eight years later the

---

[2] Ibid., 23 Dec. 1980.

Steel Corporation created BSC (Industry) as the medium through which it could help the progress of redeployment in areas in which its rationalization programmes might create major lay-offs or closures. For 18 months before the ending of production at Consett, BSC (Industry) had been working with local organizations to try to identify job-creating prospects in other industries and to attract them to the area. As soon as the close-down occurred, BSC (Industry) mounted a promotional campaign in the national press (see Plate 21.1). It also produced an attractive brochure, 'The North Country Option: Derwentside', in which Consett was characterized as 'a hardworking, no nonsense area, close knit, but warm in its welcome'. The Manpower Services Commission here, as in other steel closure areas, introduced short local training courses. The town and district were already part of the North-East Special Development Area, and therefore benefited from a high level of assisted area provision. In 1980, before the closure, the Government announced substantial financial help to the English Industrial Estates Corporation to increase provision of estates in the area (see Fig. 21.1). By late 1981, alone among districts in the Northern Region, Derwentside had been chosen by the European Community for the employment premium scheme, financed by the European Social Fund, which offered as much as 30 per cent of a new employer's wage bill for the first six months of operations. Another aid to capital investment was a system of loans pegged at an interest rate of 10 per cent from the fund of the European Coal and Steel Community. These various forms of assistance made Derwentside, in financial terms at least, perhaps the most attractive area in the whole of northern England for the development of new businesses.

Efforts were made to extend and to capitalize on the advantages with which Derwentside had been provided. Roads had been poor, there being at the time the works closed no more than half a mile of dual carriageway in the whole of the district. After 1980 a major road improvement programme was carried out. In 1981/2 Derwentside District Council commissioned a study to identify the 50 American high technology companies believed most likely to open production plants in Europe. They followed this up by sending a delegation to visit these concerns in order to 'sell' the attractions of Derwentside.

The rewards for industrial promotion exercises, for the financial assistance, and for the improvement of the environment were considerable, though the record was a mixed one (see Tables 21.3, 21.4, 21.5, and 21.6). In the first ten months of 1981 alone, 31 new companies were attracted to the area and 19 local firms expanded. The impact on unemployment was initially disappointing, even though in that year 1,600 new jobs were created. Early in 1982 there were still 8,000 people out of work in the immediate neighbourhood of Consett. Yet at a time of deep national recession the area had proved able to generate new jobs, and it seemed not unreasonable to hope that when economic recovery occurred it could do even better. By the end of 1982 there were plans for developments to produce another 1,400 jobs. In December that year the role of

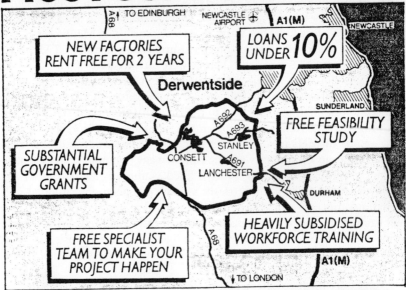

PLATE 21.1. The press advertisement of BSC (Industry) after Consett's closure

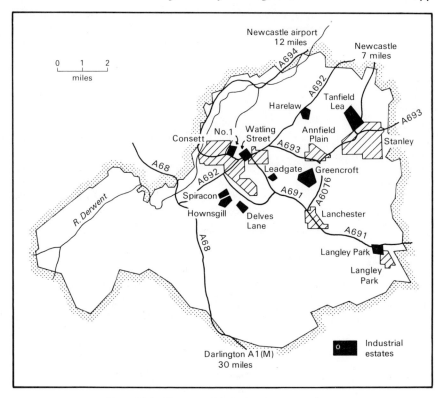

FIG. 21.1. Derwentside industrial estates, 1983

BSC (Industry) in the area was taken over by a new Derwentside Industrial Development Agency, supported by the Derwentside council, English Industrial Estates, and BSC (Industry). By 1983 there were 10 industrial estates in the area (see Fig. 21.1). By the end of summer 1984 over 150 businesses had been started in Consett with a commitment to the creation of almost 4,000 jobs, of which 2,000 were already on the ground. Sir Charles Villiers, who as BSC Chairman had recommended the closure of the works to his board, was now the head of BSC (Industry). Looking in from the outside to an area from which his company had now withdrawn, he took a generally sanguine view of the new economy: 'Some of these new businesses will rise and others fall, but there is now in Consett a much more broadly based industry than in the old days. The full regeneration of Consett will take time, but it is on its way ...'.[3] By 1988 a total of about £107 million had been spent on the economic regeneration of Consett—£41 million on the reclamation of the steelworks

[3] Sir Charles Villiers to *The Times*, 28 Sept. 1984.

TABLE 21.3. *Population 1961–1986 for districts of County Durham* ('000s)

Districts	1961	1971	1981	1986
The Central Growth Corridor				
Darlington	95.2	98.1	98.7	98.5
Sedgefield	82.8	88.8	93.5	89.4
Durham	75.0	80.1	83.7	82.3
Chester-le-Street	42.7	48.6	52.3	52.6
The East				
Easington	111.4	109.6	101.8	98.1
The Western Fringes				
Teesdale	26.5	24.3	24.1	24.2
Wear Valley	72.2	65.5	64.3	64.6
Derwentside	99.5	92.4	88.8	86.7
County Durham	605.3	607.4	607.4	596.1

Source: Durham County Council Planning Department, *County Durham in Figures* (annual).

TABLE 21.4. *Social indicators for districts of County Durham*

Districts	Death rate per '000, 1986	Rateable value per head, 1986 (£)	Households lacking both bath shower and inside w.c., 1981 (%)	Households with no car, 1981 (%)
Derwentside	14.3	90	1.5	50.7
Chester-le-Street	10.9	89	0.8	41.9
Durham	10.5	109	1.2	42.7
Sedgefield	12.5	110	1.1	48.1
Darlington	13.2	136	2.1	44.8
Wear Valley	13.3	94	2.9	49.2
Teesdale	13.0	93	2.4	33.3
Easington	13.4	81	1.3	56.7
Co. Durham	12.7	102	1.6	47.6
England and Wales	11.6	n.a.	1.4	38.5

Source: Durham County Council Planning Department.

site, on other environmental improvements, and on the building of new factories, and £16 million of public money and £50 million of private investment in about 200 new businesses. Some 3,500 new jobs had been created, and at that time it was anticipated that expansion of existing concerns and projected developments might produce 1,300 more.[4]

Through the mid-1980s the problems of Consett, Derwentside, and of the whole of north-west Durham remained serious. At the beginning of 1985 the

[4] *Financial Times*, 1 June 1988.

TABLE 21.5. *Population and unemployment in districts of County Durham, 1985/6*

Districts	Resident population, 1985 ('000s)	Male unemployment, Apr. 1986		Female unemployment, Apr. 1986	
		Total	% unemployed for more than 1 year	Total	% unemployed for more than 1 year
Derwentside	86.8	5,584	50.2	2,059	39.0
Chester-le-Street	52.2	2,496	45.4	1,011	29.9
Durham	83.1	3,166	43.7	1,438	28.0
Sedgefield	90.4	4,777	45.7	2,138	28.4
Darlington	98.9	4,547	46.8	2,045	35.9
Wear Valley	64.5	3,931	50.1	1,439	35.3
Teesdale	24.0	888	41.9	430	31.4
Easington	98.7	5,034	44.9	2,007	31.5
County Durham	599.9	30,423	46.8	12,567	32.8

Source: Durham County Council Planning Department, *County Durham in Figures* (annual).

TABLE 21.6. *Unemployment in Derwentside and in the remainder of the Newcastle and Durham travel-to-work areas in January, 1984–1988*

Year	Derwentside		Remainder of Newcastle and Durham travel-to-work areas		Derwentside as % of the remainder of the region	
	Male	Female	Male	Female	Male	Female
1984	5,768	1,998	47,531	17,476	12.1	11.4
1985	5,750	2,124	49,224	18,899	11.7	11.2
1986	5,808	2,045	50,868	19,816	11.4	10.3
1987	4,999	1,787	49,401	18,415	10.1	9.7
1988	4,451	1,527	42,251	15,589	10.5	9.8
1988 (Aug.)	3,821	1,316	36,597	13,978	10.4	9.4

Source: Durham County Planning Department, *County Durham Structure Plan* (annual report).

unemployment rate for young people in the town was 60 per cent. In April 1986, 50.2 per cent of the unemployed males and 39.0 per cent of the unemployed females in Derwentside had been out of work for over a year—for both categories the worst figures for any of the eight districts of the county. However, there had been a marginal improvement. Between January 1984 and January 1987 Derwentside's share of the male unemployment in the Newcastle and Durham travel-to-work areas fell from 10.82 per cent to 9.36 per cent, and for females from 10.25 per cent to 8.84 per cent. By August 1988 the respective figures were 9.45 per cent and 8.60 per cent. Even so, high unemployment, social deprivation, and above all that intangible but real problem of the loss of human dignity remain. During the General Election of spring 1987, when discussion of a so-called North–South divide in British economy and society was a leading issue, a Newcastle University analysis of the prosperity of 280 localities throughout Britain was made public. Consett ranked lowest of them all.

Whether or not the closure of the Consett steelworks was an economic necessity is still debatable, but once that calamity had occurred, and demolition got underway, the best efforts had to be made to rehabilitate an important community. The evidence which is provided by the history of Consett since 1980 is that British society has learned a great deal, if not yet enough, since the economic crises and the human distress of the inter-war years. Even so, this more recent history as epitomized in Consett works raises some wider issues.

22

One-Industry Communities: Private and Social Costs

THE ending of the career of Consett Iron was only one of a long series of plant closures in the history of the modern British steel industry. Even before 1900 there had been major incidents such as the run-down and abandonment of the once great works on the northern outcrop of the South Wales coalfield, and the decimation of the Black Country iron trade. Much nearer to Consett there was the closure of the Darlington Steel Company and the ending, in the first years of this century, of iron and steelmaking at Tudhoe. Later, as social consciences became keener, plant closure caused great bitterness, as, for instance, between 1920 and the mid-thirties at Dowlais, Penistone, or within the North-East, at Newburn and Jarrow. Forty years later the British Steel Corporation embarked on its own rationalization programme. In most instances this did not involve the complete closure of works: rolling mills and various finishing plant remained in operation—as at Ebbw Vale, Irlam, Shotton, and Corby. Consett's situation was that of a minority of BSC closures, which also included Bilston and Normanby Park, in which the whole of the plant was abandoned and demolished. At Consett, in contrast to both these other cases, steelmaking was located in a discrete, almost single-industry community, situated at a considerable and difficult journey from alternative employment. Its end, like its whole career, highlights the problems of the one-industry community.

Towns very heavily dependent on one industry, even on one firm, have grown up widely since the industrial revolution. They have often been related to a localized resource. This has not always been a material resource, but sometimes a special skill, reputation, or other advantage, as with shipbuilding towns such as Blyth or Greenock, railway workshop centres which afterwards became problem towns such as Horwich or Shildon, or towns dominated by motor firms such as Luton, Dunstable, or Oxford. There have also been one-industry towns oriented on a variety of types of raw material. Britain has its St Helens, Widnes, and Northwich, built up and shaped by local salt, glass sands, and coal. The Potteries region was created by the availability of local materials, though distant sources of clay early replaced the Staffordshire supplies. Iron and steel towns were a prominent group. They have been established at all stages of development of the modern industry, from Consett in 1840 to Corby 95 years later. They are indeed found throughout the world under various political regimes. The degree of dependence on the main activity varies considerably. As a community matures there is a tendency for other activities to come in to reduce the prominence of the staple trade. A great deal has depended on whether or not the local social conventions permit womenfolk to work. This in turn is related in part to the question of location and therefore to wider business opportunities. There seems no doubt that in many instances business leaders of the dominant activity discouraged new enterprises so as to avoid competition for the work-force. In Consett and district, the long-established pre-eminence of coal and steel meant a remarkable preponderance of jobs for men. In 1949, at the time when the post-war reconstruction and expansion were underway, the proportion of females in the total number of insured employees was 37.8 per cent in Newcastle upon Tyne, 38.4 per cent in Gateshead, 31.6 per cent in Blaydon, but only 18.4 per cent in Consett (Business 1951).

In modern economies, whether centrally planned, or ones in which intervention tempers the play of market forces, should such one-firm or even one-industry towns be permitted? In view of the sometimes calamitous results of their failures, there seems to be a strong case that they should not. But how can they be avoided? In a free market economy with a *laissez-faire* minimum of constraints on enterprise, this cannot be easy. In either a centrally planned economy or a mixed economy it may be that planning permission (physical planning involving zoning of land for various uses, and economic planning, perhaps using some control device such as the industrial development certificates which were long required in Britain) might help towards a solution of this problem. It is claimed that under socialism the one-industry community is not at risk. If that is so then it must pose questions about the rationality of socialist location policies which are at least as serious as those which are raised by the closure programmes in British basic industries.

What then should happen when the staple trade of a large, one-industry town is judged to be no longer viable? Even if 'unviable' means, narrowly interpreted, that a works is no longer economic as a production unit, there may be possibilities of redeeming the situation by diversifying into new, often higher-value products in the same broad line of business. If this is considered impossible, or is tried and fails, so that the plant becomes surplus to the requirements of the markets for its products, should it then be kept open merely to provide jobs and the support for a local community? It may be argued, as suggested above, that the calculation of the true cost to the nation of closure involves not only balancing the writing-off of the plant and its equipment against the reduction of national production losses (if any), but also the cost of the destruction of the social overhead capital in the once dependent and now possibly blighted community. However, the question must also be asked, 'is this a concern which can long be allowed to occupy the centre of the stage in a competitive world?' The situation in which such social concerns assume great significance in decision-making may in fact amount to a continuing subsidizing of 'lame ducks' by more efficient parts of the economy. This spreads the costs of maintaining old, less efficient structures and locational patterns of manufacturing, but eventually, through reducing the general level of competitiveness, it may impoverish the whole community. Such a policy may require the country which pursues it to cut itself off from world trade and competition to a greater or lesser extent. Given its dependence on overseas sources of supply for a large part of its raw materials, and especially for food, it is doubtful whether such an option is an acceptable one for Britain. That is to say that her inextricable involvement in the world trading system, Britain's great strength in her nineteenth-century heyday, has become her mid- and late twentieth-century trap. With her greatly increased population and her enhanced material expectations she must continue to dance to the tune which a competitive world now plays, and yet of which she began the performance. In these circumstances there is every reason to anticipate that the experiences of Consett will be repeated time and again in many industries. This has indeed already happened on a number of occasions since 1980, though rarely with the same dramatic setting.

In coming to such conclusions one courts the possibility of being accused of a lack of social concern or of compassion. That all life is not economic is a truism which capitalism has too often denied by the narrowness of its theory and still more by the meanness of its practice. The new prominence of social and recreational life, as both essential to the worker and his inalienable right, points up the fact that work is only undertaken to satisfy human needs. Yet it remains true that production is a commercial activity and should be efficiently conducted according to the principles appropriate to that sector of life. The conclusion must be that a healthy society as well as a healthy economy must move with the times, must be willing to close old plants and to redistribute its production, and to that end has to redeploy and redistribute its industrial population. But if that is so, then the utmost efforts must be made to find the displaced work-force alternative employment. If workers have to move, their problems with housing must be dealt with imaginatively and sympathetically, and the impact on those who remain of the removal of the most enterprising by out-migration must be taken into the account. It should be incumbent on government to face up to the fact that it is probably no more than ignorance or prejudice that today causes entrepreneurs in new-style, growth industries to fail to recognize that under contemporary conditions of manufacturing, increasingly footloose in character, the Rhondda may be as suitable a location as Reading, and Consett as favourable as Chelmsford. On the other hand, if replacement jobs after a major plant closure cannot be found either in the area directly affected or elsewhere in the nation, can it still be regarded as a legitimate object of national policy to maintain a low level of unemployment in Britain? In such circumstances it can perhaps be maintained that the nation is substantially over-populated. Even in a consumer-oriented society, the loss of demand from a reduced population would be less serious than the gain in reduced unemployment. More important still, in default of the continuance of the near full-employment policies, regarded as so central to the concept of the post-war welfare state, the very essence of British democracy, as more than one generation has come to understand it, is at stake. For those areas and those individuals losing their livelihood, and receiving nothing to replace it, the continuing provision of basic amenities and distractions of various types instead of work may be merely the modern counterpart of bread and circuses: video films and cable television may be the modern opium of the people.

Reflections on the closure of Consett works and its aftermath raise a host of wider questions concerning not only the future of manufacturing in Britain, but also her value systems and social and political organization. The longer-term history of Consett highlights other, if narrower and more prosaic, issues. It points to the possibility for a long and successful career even for a poor industrial location so long as the business is pursued with imagination and vigour. To that extent Consett's career is a challenge to the wisdom if not to the logic of those who continue to pursue a study of industrial location which is concerned with 'optimum' choices. This plant's commercial success owed most to the leadership at the top of the firm, but much also to the complementary abilities and loyalty in the ranks. Its existence created a rich society, not so much in material terms, for such benefits have all too often been creamed off to the advantage of the few, but in the wider, human values. The Consett story also shows that there will be periodic crises during which the survival of a firm becomes difficult. Such occasions are however also opportunities for growth. Paradoxically such crises have proved more difficult to overcome in an industry organized nationally than when the fight was conducted by a single beleaguered firm. If Consett had remained in private hands in 1967, or had passed to the private sector consortium in the dark days of September 1980, it might well have survived. Yet it remains true that, had that happened, it would have been a case of the good, even of the commercially viable, frustrating the better, or at least the potentially better. Is such a situation a full enough justification for the existence of a nationalized industry, which, in principle, should be able to take the broader view? Looking at the development of Consett he would be a bold man who would make such a claim.

Locational change has happened on a massive scale throughout the history of the British iron and steel industry. When on a June morning in 1850 John Marley and John Vaughan stumbled on the outcrop of the Cleveland Main Seam on the hillside near Middlesbrough the economic rationale was removed for the major ironworks at Consett, though that plant was then still not 10 years old. The fact that it survived for another 130 years is a remarkable testimony to the permissiveness of locational factors. It is also a tribute to those who worked in that company, generation after generation. Consett works produced not only iron and steel, for some considerable wealth, and for others hard labour and even drudgery, but also in the process some remarkable men

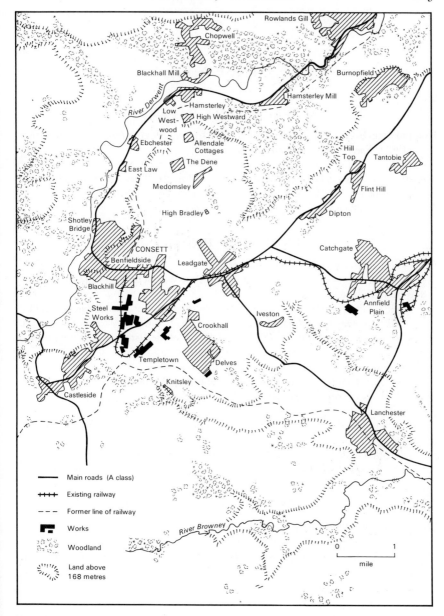

FIG. 22.1. Consett and neighbouring parts of north-west County Durham, c.1980

and a distinctive community. In these ways, in spite of what has happened in recent years, the course of Consett Iron's long and distinguished career is still stamped firmly on the landscape of north-west Durham (see Fig. 22.1).

REFERENCES

GOVERNMENT PAPERS

Iron and Steel Board (1957), *Development in the Iron and Steel Industry: Special Report.*
—— (1961), *Special Report on Development.*
—— (1964), *Special Report.*
'Report of the Departmental Committee Appointed by the Board of Trade to Consider the Position of the Iron and Steel Trades after the War' (1918), Cd. 9071.
Royal Commission on Coal Supplies (1904), evidence of J. S. Jeans, BPP 1904 XXIII.

INDUSTRY PAPERS

BISF (1941), *Jarrow and the Steel Industry.*
—— (1944/5), 'Notes by S. L. Bengston on the Post War Schemes of the British Iron and Steel Industry' (unpublished).
—— (1945), *Report to the Minister of Supply on the Iron and Steel Industry* (Dec.).
—— (1946), *Comparison of Capital and Steel Output of North-East Coast Steelmakers.*
—— (1966), *The Steel Industry: The Stage I Report of the Development Co-ordinating Committee of the British Iron and Steel Federation.*
BITA (1878), *Report on the Iron, Steel and Allied Trades.*
BSC (1971), 'Location of Greenfield Site' (June, unpublished).
—— (1973), *Ten Year Development Strategy.*
—— (1979a), *Iron and Steel Works Plant Census.*
—— (1979b), *The Return to Financial Viability: A Business Proposal for 1980/81.*
—— (1980a), *Consett: Case for Closure* (June).
—— (1980b), *The Case for Consett Closure: Supplementary Information.*
—— (1980c), *Appraisal of the Trades Union Brochure 'No Case for Closure'* (Aug.).

TRADES UNION DOCUMENTS

Consett Unions (1980), Joint Trades Unions at Consett Works, *No Case for Closure: The Trade Union Alternative Strategy for Consett Works* (July).
ISTC (1977), *Plate Mill Development: A Critique of the Revised BSC Submissions of 1977.*
—— (1980), *New Deal for Steel.*

GENERAL

ABERCONWAY, Lord (1927), *Basic Industries of Great Britain* (London).
ALDCROFT, D. (1964), 'The Entrepreneur and the British Economy, 1870–1914', *Economic History Review*, pp. 113–34.
ATKINSON, J. R. (1968), *The Steel Industry in North West Durham*, a study for Durham County Council (Durham).

References

Bell Papers (n.d.), Papers relating to mining, Newcastle Central Library.
BELL, I. L. (1863), *Report on the Iron Trade of the North East*, Report of British Association, Newcastle meeting (London).
BELL, T. H. (1883), Paper to the Iron and Steel Institute reported in *Engin.* 28 Sept.
BILLY and MELIUS (1904), quoted in *Iron Age*, 22 Sept.
BIRD, C. (1881), *A short Sketch of the Geology of Yorkshire* (London).
Birmingham (1857), Pamphlet on 'Iron', in Birmingham Public Library.
Board of Trade (1932), *An Industrial Survey of the North East Coast Area* (London).
BOSWELL, J. S. (1983), *Business Policies in the Making: Three Steel Companies Compared* (London).
British Association (1970), *British Association Handbook: Durham County and City with Teesside* (Durham).
BRUCE-GARDNER, C. (1930), 'Report on the Structure of the Iron and Steel Industry of Great Britain' (31 Dec.; unpublished report for the Bank of England).
BRYER, R. A., BRIGNALL, T. J., and MAUNDER, A. R. (1982), *Accounting for British Steel* (Aldershot).
BURN, D. L. (1940), *The Economic History of Steelmaking, 1867–1939* (Cambridge).
BURNHAM, T. H., and HOSKINS, G. O. (1943), *Iron and Steel in Britain, 1870–1930* (London).
Business (1951), Business Publishers, *Market Survey of the United Kingdom* (London).
CAMPBELL, J. G. (1958), conversation with the writer.
CARNEY, J. (1980), *Economic Audit of Consett Steelworks* (June).
CARR, J. C. and TAPLIN, W. (1962), *History of the British Steel Industry* (Oxford).
Consett Records (British Steel NRRC 1027).
Consett (1893), *Consett Iron Co.: Description of the Works* (Consett).
COOKSON, R. (1980), conversation with the author.
COOPERS and LYBRAND (1979), *Derwentside: A Strategy for Industrial Regeneration* (London).
CRONIN, J., and TROW, E. (1892), *Royal Commission on Labour*, Vol. 2.
DALE, D., David Dale papers (British Steel NRRC 1027/3/31–2).
Derwent Papers (1857, 1858), Papers on failure of Derwent Iron Company, in Durham County Records.
Derwentside District, Annual Job Audit and miscellaneous papers.
Dowlais (1859), 'Report by E. Williams on North and Midland Ironworks', Dowlais Company Records.
Durham County Council (1981), *County Structure Plan*, and subsequent annual reports.
ELLIOTT, C. (1982), 'Consett' (unpublished dissertation of Lanchester College, Coventry).
English Cyclopaedia (1854), Bradbury and Evans (London).
European Iron Ore Co. Ltd. (1938), 'Report to the British Iron & Steel Federation on the Iron Ore Discharging Facilities of the Ports of the United Kingdom' (London).
FISK, T. A., and JONES, T. K. (1971), 'Regional Planning from the Point of View of a Major Employer', *Regional Studies Association Conference on Regional Development* (Sept.).
FORDYCE, W. (1857), *The History and Antiquities of the County Palatine of Durham* (Newcastle).
—— (1860), *History of Coal and Coalfields* (London).

HABAKKUK, H. J. (1962), *American and British Technology in the Nineteenth Century* (Cambridge).
HALL, W. N. (1865), 'Information Related to the Production of Coke in Northumberland and Durham' (Durham County Archives).
HARBORD, F. W. (1927), Presidential Address to the Iron and Steel Institute, *Iron and Coal Trades Review*, 6 May 1927, p. 711.
HODGKIN, J. E. (1913), *Durham* (Methuen's Guides; London).
HUDSON, R., and SADLER, D. (1983), 'Anatomy of a Disaster: The Closure of Consett Steelworks', *Northern Economic Review*, 6.
Iron and Steel Institute (1871), List of British Blast Furnaces (Aug.).
Iron and Steel Industry Annual Statistics, United Kingdom Iron and Steel Statistics Bureau (London).
JEANS, J. S. (1875), *Pioneers of the Cleveland Iron Trade* (Middlesbrough).
JISI (1870), 'Recent Developments in the Iron Trade'.
LANDES, D. (1969), *The Unbound Prometheus: Technological Change and Industrial Development in Western Europe from 1750 to the Present* (Cambridge).
LAYTON, W. T. (1922), 'The Influence of the War on the British Iron and Steel Industry', *Manchester Guardian Commercial*, Sept. 1922.
LEWIS (1848), *Topographical Dictionary of England* (London).
LOUIS, H. (1916), *British Association Handbook*, Newcastle upon Tyne meeting (Newcastle).
MCCLOSKEY, D. N. (1973), *Economic Maturity and Entrepreneurial Decline: British Iron and Steel, 1870–1913* (Cambridge, Mass.).
Metal Bulletin (1974), *Iron and Steel Works of the World* (London).
MILLS, H. R. (1962), (of British Iron and Steel Research Association) quoted in *Steel and Coal*, 15 June, p. 1127.
Mineral Statistics of the United Kingdom (annual).
MOTT, R. A. (1936), *History of Coke Making* (Cambridge).
MOUNTFORD, C. E., et al. (1975), *Langley Park Colliery, 1875–1975* (Chester-le-Street).
NAVE, G. M. (1961), conversation with the writer, Aug.
NEASHAM, G. (1882), *The History and Biography of West Durham*, quoted in *Colliery Guardian*, 5 May, pp. 707–8.
Newcastle (1887), *Royal Mining Engineering and Industrial Exhibition* (catalogue).
PARSON and WHITE (1828), *History and Directory and Gazetteer of the Counties of Durham and Northumberland* (Newcastle).
Penny (1833–43), *Penny Cyclopaedia* (London).
PEP (1933), *Report on the British Iron and Steel Industry* (July).
Planning and Development (1982), 'Report to the Planning and Development Committee of Derwentside District Council' (Apr.).
PORTER, G. R. (1846), 'On the Progress, Present Amount and Probable Future of the Iron Manufacture in Great Britain', *British Association, Southampton: Report*.
PRICE, K., and WADE, E. (1983), *A Town in County Durham: A Community Study of Unemployment* (June, Open University, Northern Region).
RICHARDSON, H. W., and BASS, J. W. (1965), 'The Profitability of the Consett Iron Company before 1914', *Business History*, vii. 2. 71–93.
RICHARDSON, T., and WALBANK, A. (1911), *Profits and Wages in the British Coal Trade, 1898–1910* (Newcastle).
ROBERTS, P. W. (1970), 'Consett Study, 1969–70' (unpublished MA dissertation, Newcastle University).

Rylands (various), *Rylands Directory of the Coal, Steel and Allied Trades* (Birmingham).

Slaters (1877), *Directory of Northumberland, Durham & the Cleveland District* (Manchester).

SMT (1932), "Securities Management Trust, Scheme of Reorganisation of the Steel Industry in the West of Scotland" (Sept., Bank of England).

THACKEREY and LOCKLEY (c.1890), 'Report on East Coast Mills to E. P. Martin', Dowlais Iron Co. Records, Cardiff, Box 1.

TOLLIDAY, S. R. (1987), *Business, Banking and Politics: The Case of British Steel, 1918-1939* (Cambridge).

TOMLINSON, W. W. (1914), *The North Eastern Railway: Its Rise and Development* (Newcastle).

UPHAM, M. (1988), Research Officer of the Iron and Steel Trades Confederation, in correspondence.

VAIZEY, J. (1974), *The History of British Steel* (London).

WARREN, K. (1969), 'Recent Changes in the Geographical Location of the British Steel Industry', *Geographical Journal*.

WATKINS, D. (1968), Two articles on 'The Future of Consett', in *Voice of North East Industry*, Sept. and Nov.

WATTS, P. (1902), 'Ships and Shipbuilding', in *Encyclopaedia Britannica*, 10th edn. (London).

WHELLAN (1894), *Directory of County Durham* (London).

WILKINSON, E. (1939), *The Town that was Murdered: The Life Story of Jarrow* (London).

WILLIAMS, E. (1869a), 'Report by Edward Williams on Consett Works', dated Middlesbrough, 8 Mar., CDM 13/3/69.

—— (1869b), 'Second Report on Consett Operations', CDM, 8/6/1869.

WILLIS, W. G. (1969), *South Durham Steel and Iron Co. Ltd.* (Hartlepool).

WRIGHT, C. R. A. (1880), 'Iron', in *Encyclopaedia Britannica*, 9th edn.

INDEX

Aberconway (MacLaren), comments on Consett profitability, 1905 and 1927 76, 89
Ainsworth, G.
  assessment of Derwent Haugh in 1890s 60–4
  as General Manager 70, 71, 80
  tour of American works 71
amalgamation proposals 93, 94, 100
angle mills, built 1888–93 51
Atha, C. G., evidence to Iron and Steel Industries Committee (1917) 85
Attwood, C., and iron developments in County Durham 12, 30

Bankers Industrial Development Company, and Jarrow 112
basic Bessemer process, and plans for Jarrow steelworks in 1930s 109
Bessemer process, compared with puddling 44
Bengston, S. L.
  and Consett's 1930s development policy 107
  and Jarrow works scheme (1937) 112
  on 1944/5 plans for the North East 115, 117, 118
Benson Committee
  Consett Iron reactions to 141–4
  Report (1966) 141
Beswick, Lord, review of plate mill schemes (1975) 153
Bilbao ore 42
Billets
  markets and production in late 1970s 162–8
  mill installed at Consett (1950–3) 119
Billy and Melius, comments on British inefficiency (1904) 88
Bishopwearmouth works 25, 28
Bolckow Vaughan and Company
  in early 1850s 11–13
  in Great War 84
  1929 merger with Dorman Long 100
  steel plans in 1870s 46
Brasserts
  and Consett relocation in 1920s 99
  on Jarrow 110
  1936 report on Consett 107
British Iron and Steel Federation
  development co-ordination introduced (1936) 103
  established (1934) 102
  1945 Development Plan 118, 119

British Steel Corporation
  stages in development thinking (1967–80) 145, 149–51
  Ten Year Development Strategy (1972) 150, 151
British Steel (Industry), and Consett rehabilitation 174–7
Bruce-Gardner, C.
  and Jarrow 112, 113
  on North East coast reconstruction in early 1930s 103–5
Bryer, Brignall, and Maunder, on financial aspects of BSC rationalization programme 150, 166

calcining of ore 11, 38
Cargo Fleet works
  established, 71
  molten iron from, to Consett (1969) 148
  post-World War II plans for 117, 118
Cartwright, F, on British ore movement in 1960s 140
Chopwell, coal developments 57, 72
Cleveland ore
  Consett operations and interest in 12–14, 42
  discovery of Main Seam 11
  early use of in North East 9
closure
  anticipated impact and strategy 155
  arangements (1980) 158
  effects (1980 onwards) 171
coal
  and coke in Consett success 76
  developments by Consett Iron 37, 38, 71, 72, 89, 90
  in north west Durham in post World War II period 139, 145, 147, 171
coal measure iron ore 5, 13
coke
  rate of consumption in iron smelting (1850s to 1870s) 45
  rate post-World War II, 138
coking practice
  in 1920s 89, 90
  modernization on Teesside in 1930s 105, 106
Colvilles, 1950s plate mill schemes 128
Consett Iron Company
  capacity (1974), (1979) 154

Index

capacity by departments: (1894–1912) 68, 71, 77; (1922) 92
  equipment in late 1970s 160, 161
  material and intermediate product flows, 1950s 127
Consett town
  dependence on ironworks 57
  establishment and growth 52–8
Cookson, R.
  on Benson Report (1966) 141
  comments on Kaldo process 136
  on co-operative rolling agreement with Dorman Long (1964) 134
  criticism of closure proposal 157
costs of production
  in 1860s 14, 30–35
  Consett and other British works (1978–9) 161
Cumbrian iron ore
  at Consett in 1860s 15, 41
  survey of in 1930s 107

Dale, D. 23, 25, 81
Darlington Iron Company
  failure 70
  finished iron capacity 47
Departmental Committee of the Board of Trade on Steel (1917) 85–7
depression of trade
  in 1870s 46–8
  in early 1920s 94
  from 1929 101, 102
Derwent and Consett Iron Company (1858) 23
Derwent Haugh
  consideration for ore terminal (1948) 115
  as possible location for iron and steel making (1896–9) 60–5
Dorman Long
  Redcar project (1916) 84, 92
  co-operative plate rolling arrangement with Consett (1964) 134
  new coke ovens (1935) 106
  plate mill schemes in 1950s 128
  postwar situation 117
  takes over Bolckow Vaughan (1929) 100
  1930 discussions with Consett 103–5
dividends and shares of North East steel firms
  (1883) 75
  (1900–12) 69
  (1920) 97

efficiency and cost reduction
  early Consett stress on 9, 10, 28, Ch. 11, 96
  national, interwar 101
employment in north west Durham (1951–76) 146
entrepreneurial failure and British industry in the late Victorian age 1, 2, 67

Fell Coke works
  pioneering role of 89, 90
  post-World War II 117
  small scale of by 1970s 160
finished iron
  collapse 46, 47
  early 15
  in 1860s and early 1870s 44
Ford Motor Company, consideration of Washington for motor plant (1960) 133
foreign competition, in early 1920s 94
Franks Report (1945) 114
Frodingham Iron Company, competition in angles 70

George, E. J., 1923 comments on rail freights and Consett rebuilding 99, 100
Grunner and Lan, assessments of Consett operations (c.1860) 14, 15

Harbord, F. W., comments on plant modernisation (1927) 98
Hareshaw ironworks, Northumberland 6, 15
Hownsgill plate mill
  closure (1979) 154
  commissioned (1960) 131
  deficiencies and BSC plans for a new wide mill 153

Import Duties Advisory Committee 102
Industrial estates in Derwentside in 1980s 177
Iron and Steel Board
  on 'obsolescence of location' 137
  and plate proposals in 1950s 128–31
Iron and Steel Holding and Rehabilitation Agency, offer of Consett Iron for sale (1955) 122
Iron and Steel Trades Confederation
  on BSC plans generally in late 1970s 156
  struggle to save Consett works 165–8
iron industry of North-East
  in early nineteenth century 5
  in 1840s 8, 13
  (1840–80) 16, 31
  (1892) 50
iron ore
  deficiencies in North East before 1850 5, 6
  deliveries and consumption at Consett in 1860s 39

Japan, as model for BSC reconstruction in late 1960s and in 1970s 149
Jarrow
  proposed basic Bessemer steelworks in 1930s 109
  works of Palmers 102, Ch. 15
Jenkins, W.
  appointed General Manager 35, 36

Index

Jenkins, W. (*cont.*):
    efficiency in management 80, 81
    and local society 55
    retirement and death 70

Kaldo process, trials at Consett 136

Labour and wages 78–80
Langley Park mines and coke ovens 38, 57
Layton, W. T., comments on Great War extensions in British steel 87
location
    change in conditions following Cleveland ore discovery 11
    social v. economic considerations Ch. 22
    theory 2–4, 184

McCloskey, D., views on *c*.1900 efficiency of British steel industry 2, 67, 68
markets for plates and billets by region (1965 and 1980) 163
migration of skilled ironworkers to North-East 53

National Ports Council, ore dock plans (1965) 140
National Shipbuilders Security Ltd., and Jarrow works 109, 113
Newburn steelworks 100, 102, 108
New Jarrow Steel Company
    becomes Consett Jarrow Rolling Mills (1948) 113
    formed (1938) 113
North-East Coast, output by firms
    (1937) 107
    (1937–50) 115
    (1937–84) 169
North-East Coast Steelmakers Association 79, 96
North Eastern Railway
    credits to Derwent and Consett Iron Company in early 1860s 25
    formation and significance 20
Northern Industrial Group, 1980 plan to revive Consett operations 168–70
Northumberland and Durham District Bank, crisis associated with failure of (1857/8) 22–5
Nuffield Trust, and Jarrow 111, 112

Orconera Iron Ore Company (formed 1873) 42
ore docks
    conditions in 1950s 138
    developments in 1960s 140
    1917/1918 proposals for British steel industry 86, 87
Oxygen (L. D.) steelmaking
    at Consett 131, 136
    extensions in 1968 146

paternalism of Consett Iron Company in nineteenth century 53, 55
Pearson, S. C., statements on Consett production costs (1966/7) 142, 145, 146
Pease, J., comments on situation of Derwent Iron Company (1857/8) 22, 23
plate production
    BSC forecast (1971) 151
    deliveries (1971–80) 153
    developments in 1950s 122, 128, 129, Ch. 17
    efficiency of new mills in 1920s 96
    Great War extensions in United Kingdom capacity 84
    iron 28
    in late 1870s and early 1880s 48
    national output by efficiency categories (1955–65) 132
    output (1952–69) 133
    steel plate 48–51
Political and Economic Planning (PEP), report on steel industry (1933) 101
population
    of Conside cum Knitsley (1841–61) 53
    of Derwentside district (1951–86) 172
    and social indicators for districts in 1960s and 1970s 178, 179
Portal, Lord, and Jarrow project 111, 112
price control and development prospects in the steel industry in 1960s 136
Priestman family
    coal interests north of the Derwent 72
    and Consett 23, 35
productivity in Consett operations in late nineteenth century 78, 79
profits
    structure of: (1870) 45; (1870–87) 50; (1899–1922) 89; (1956–66) 134

railway freight rates and Consett
    in 1850s 15
    in 1860s 39–42
    after 1870 75
    in 1890s 65
    in 1920s and 1930s 99, 107
rate burden in 1920s 99
rationalisation planning in 1930s 102, 103
Redcar
    ore terminal announced (1969) 148
    plate mill plan: (early 1960s) 134; (1972–6) 153
redundancy payments at Consett 168, 173
Richardson and Company, Shotley Bridge works, purchased 25
Richardson, W. 1839 initiation of mineral surveys leading to founding of Consett works 6

Index

Ridsdale ironworks, Northumberland 6, 15
Rosedale ore, for Consett c.1860 15

Sahlin, A., evidence to Iron and Steel Industries Committee (1917) 85
sea transport of iron ore
 arrangements for Bilbao ore in 1870s 42
 in plans for redevelopment of British steel after Great War 86, 87
shipbuilding
 launchings (1952–69) 133
 as market: (1875–90) 49; (1880–1913) 68
 North East and other districts (1899–1920) 78
 and ship plate prices (1913–29) 93
Skinningrove works in early 1930s 105
Smith, C., chairman's comments on Consett reconstruction 93, 95
social considerations
 in closure of Consett 173, 174
 in location, 3, 4, Ch. 8, 108, 109
South Durham Steel and Iron Company
 capacity for large diameter pipes 133, 134
 development plans in Great War 84
 old technology installed at Greatham works 131
 plate mill scheme in 1950s 129
 proposed merger with Dorman Long in early 1930s 100, 103
 after World War II 117–19
Stanhope and Tyne Railway 6, 18
steel
 advance in production and markets for in 1880s 49
 Consett begins production 49–51
 early uses of bulk steel 45
Stewarts and Lloyds, and development plans for Teesside (1966) 134, 135
strike in British steel industry (1979/80) 156
Stockton and Darlington Railway
 extension of network 18, 20
 in relation to Consett operations 21

Talbot, B.
 1930 discussions with Consett on amalgamation or working arrangements 103
 on postwar development plan and South Durham Steel and Iron 118, 119
Talbot, C, character and role in plate mill expansion of 1950s 130–5
tariff protection in interwar years 102
Teesside
 companies compared with Consett (1953–81) 159
 competitive position of works in late nineteenth century 81
 first works 11
 rationalization in interwar years 100
Tow Law works 11, 12
Tudhoe
 competitiveness and relocation 70, 71
 steelmaking at 45
Tyne dock
 improvements in 1930s 107
 inadequacies recognized after nationalization and closure of ore dock 148
 limitations of, and expansion scheme for in 1960s 138, 142
 new Consett ore terminal, post World War II 115–17
 opened (1859) 59

unemployment
 in north-west Durham in 1980s Ch. 21
 of puddlers with collapse of wrought iron industry in 1870s 46, 47
United Steel Companies
 Appleby plate mill project 84, 92
 possible Consett association with in 1920s 93
Upleatham iron ore 12, 14

Watkins, D.
 anticipations of impact of Consett closure 155, 171
 on ore handling in the North East 148
Weardale iron ores 5, 9
West Durham collieries, ironworks and railways (1862) 19
Williams, E., assessments of Consett efficiency and development proposals (1869) Ch. 5
Witton Park works 12
wrought iron v. steel 46, 47